KT-177-823

TOPICS IN APPLIED GEOGRAPHY

TOPICS IN APPLIED GEOGRAPHY
edited by Donald Davidson and John Dawson

titles published and in preparation
Slum housing and residential renewal
Soil erosion
Human adjustment to the flood hazard
Office location and public policy
Vegetation productivity
Government and agriculture
Soils and land use planning
Pipelines and permafrost

Gareth Jones
University of Strathclyde
Glasgow

VEGETATION PRODUCTIVITY

Longman
London
and New York

Longman Group Limited London

Associated companies, branches and representatives throughout the world

Published in the United States of America by Longman Inc., New York

© Gareth Jones 1979

All rights reserved. No part of this publication may be reproduced, stored in a retrieval system, or transmitted in any form or by any means, electronic, mechanical, photocopying, recording, or otherwise, without the prior permission of the Copyright owner.

First published 1979

British Library Cataloguing in Publication Data

Jones, Gareth
Vegetation productivity. – (Topics in applied geography).
1. Phytogeography 2. Primary productivity (Biology)
I. Title II. Series
581.9 QK101 78-40985

ISBN 0-582-48577-0

Printed in Great Britain by
Richard Clay (The Chaucer Press) Ltd, Bungay, Suffolk

CONTENTS

LIST OF FIGURES

(*sources are cited on the Figures*)

LIST OF TABLES

(*sources are cited on the Tables*)

INTRODUCTION

The study of vegetation is an immense one. Few biogeographers or ecologists have the time, or the ability or indeed the interest to study all the aspects of vegetation patterns on this planet. For this reason 'specialists' have emerged, each specialist being responsible for a particular aspect of the overall investigation of the plant cover of this planet. Thus we have the plant phytosociologist who is concerned with the detection of plant communities and plant societies within the total vegetation assemblage. The bio-statistician attempts to pick out pattern from a seemingly totally random distribution of vegetation; this task has been assisted by the development of suitable statistical methods. Yet other specialist areas have developed around the mapping of vegetation and the determination of environmental variables which can control distribution patterns. Then there is the 'laboratory approach', where plants are grown in controlled greenhouse environments in an attempt to discover maximum growth potential, or the significance of trace elements or the role of temperature, water, or carbon dioxide availability on photosynthesis. The list is endless. Research into one problem serves to stimulate research into another area.

In all probability man has been concerned with vegetation patterns since earliest times. To Neolithic man the distribution of forests and open land was a matter of major concern. In the forests there lurked his enemies – some very real such as the bear and the wolf, some no doubt imaginary such as devils and fearsome creatures – for the forest would have been a darkened, tangled world of trees and undergrowth. We have no written record of early man's reaction to vegetation patterns, but we have inferential evidence. The creation of forest clearings by burning and felling occurred early in man's evolution in Europe, probably about 3000 BC (Renfrew, 1977). By the time of the early Greek civilization (400–300 BC) exploration, discovery and scientific reasoning were all sufficiently advanced to allow Theophrastus (371–286 BC) to postulate a world composed of distinct vegetation zones. These zones all coincided with specific latitudinal belts of distinct climatic types. Thus at the equator there was a Megatherm zone wherein vegetation development was most intense. On either size of this zone was the Xerophyte zone, an area deficient in moisture and the location of the great deserts. Beyond again were the Mesotherm areas which corresponded to the Mediterranean regions; while approaching the poles were the Microtherm areas; and finally the Hekistotherm areas, or what we call today the Arctic/Antarctic regions.

The boundaries of these regions were crudely located but credit must be given to the early Greek philosophers for appreciating that a macro-zoning of vegetation did occur. It was not until AD 1805 when the famous German geographer Alexander Von Humboldt published his *Essay on the Geography of Plants* that a significant advance was made. Von Humboldt recognised sixteen separate vegetation regions. Some years later the French botanist de Candolle (1855) published his work showing 24 vegetation regions. As the volume of knowledge increased then so greater refinement of ideas was possible. Both Von Humboldt and de Candolle recognised some anomalous vegetation units – vegetation which seemed out of sequence or influenced by factors which could override the significance of climatic control. The main area was to be found within the North American continent where the American Indian had for generations been using fire both to maintain and extend grassland.

We have come a long way since these early introductory studies on vegetation distributions. We know that climate has an overriding control on vegetation patterns and that this is due mainly to the inputs of solar radiation and moisture. The relative abundance of these two prime variables will determine the ability or otherwise for vegetation to grow. Macro-variations in the radiation and moisture variables provide us with our broad global bio-climatic regions (*see* Fig. 2.1). These are the Zonal vegetation groups. Within the zonal areas can be found examples of vegetation which appear to contradict the zonal type. These are the Azonal vegetation units which have developed in response to local factors such as local climatic patterns, the nature of the parent rock, age of the soil, relief and slope patterns and the effectiveness of man as a modifier of vegetation (the anthropogenic factor).

It is perhaps the anthropogenic factor which is now the most relevant local control over vegetation development. Indeed over about 40 per cent of the land surface of the planet man has superimposed his own vegetation patterns in the form of organised land-use, mainly agriculture and forestry. Now, more than ever before in the history of man on this planet, we are beset by the daily problems of finding adequate food and raw material supplies. To those of us living in a highly technologic society that statement may appear an exaggeration of fact, but it is not. The cost of basic organic raw materials has risen rapidly over the last five years. From figures compiled by the Department of Employment and published in Index of Retail Prices the price of sugar in Britain increased from an average of £70 per ton in 1972 to £145 per ton in 1976. Cocoa over a similar period rose from £250 per ton to £1,250 per ton and is still rising fast. Both these examples represent extreme price increases. Most agricultural commodities – feedstuffs, fertilizers, implements – have all shown dramatic price rises over the past two years and these increased costs are passed on to the consumer. In addition world population continues its relentless increase and more people demand more food while developed nations strive to improve still further their quality of life. There is little doubt that there remain vast areas of land on which agricultural production could be increased if only political and economic factors were different. With an estimated world population of 4 000 million in 1976 (Ehrlich and Ehrlich, 1970) and increasing at 55 million per annum (Huxley, 1970), the quest for organic resources will become more intense. We can but assume that the cost of food in industrialised countries will rise and that the incidence of starvation in countries with a poor technologic base will increase.

This is the background against which the need for accurate assessments of vegetation productivity must be set. The main concern of this book is to examine the different methods whereby vegetation productivity can be measured. The early chapters are devoted to the explanation of growth processes and mechanisms in plants and to the world patterns of vegetation productivity. Treatment of these themes is, by necessity, theoretical in nature but they help provide a base from which to understand man's use of vegetation resources. The approach of the book becomes increasingly applied in that it seeks to examine the ways in which man uses vegetation production. Separate sections are devoted to forestry, agricultural crops and grassland while a specific case study of a highly managed forest site is given. The final chapter draws together the wide ranging problems which have been examined earlier in the book and an attempt is made to project our knowledge of vegetation production towards the turn of the century. Questions are posed: some can be answered, while others are clouded with uncertainty and hence remain unsolved. In the future it is likely that every small piece of biotic production will be of value and no part of our planet should be producing at a below-optimum level. But who should be responsible for deciding where the optimum level of productivity should be placed? What productivity/production measurements should be adopted? Perhaps an even more fundamental question would be: 'Should we allow our life patterns to be dominated by organic productivity values?' The implications behind these questions along with the responsibility of providing answers are immense but the challenge is there and biogeographers will be expected to provide some of the answers.

The study of vegetation productivity represents just a small part of the total debate of man's impact upon the biosphere. The content of this book should not be read in isolation from other aspects of biogeography nor indeed from geographic and biological literature in general. While the topic of vegetation productivity in itself is a precise study, the implications and applications of its results are not.

During the preparation of this book I enjoyed many interesting discussions with fellow colleagues and students; their interpretations of ideas and results have greatly helped my own progress through the manuscript. Particular thanks goes out to all those persons who knowingly or unwittingly helped my ideas progress from simple concepts through to the appearance of words upon a page. To Dr Donald Davidson my particular thanks for answering many queries concerning the technical requirements of such a book and for his patience in reading the entire manuscript. My typist, Mrs MacLeod, has proved of great help, deciphering my handwriting and ensuring that I kept to deadlines while all the figures were ably prepared by Mrs Marion Walker. My final acknowledgement must go to the family; my wife, Lucy, and three children have borne the brunt of my inconsistent temper during the good and bad phases of writing. To them all, a sincere thank you.

Gareth Jones
Glasgow
September 1978

ACKNOWLEDGEMENTS

We are grateful to the following for permission to reproduce copyright material:

American Geographical Society for a figure from an article 'Geographical Aspects of Productivity', by N. I. Bazilivich *et al.,* from *Soviet Geography,* 12, 1971, reproduced with permission of the American Geographical Society; Blackwell Scientific Publications Ltd., for a figure from *Methods for Estimating the Primary Production of Forests,* International Biological Programme Handbook No. 2; Blackwell Scientific Publications Ltd., and John Wiley and Sons Inc., for our Figure 1.4 compiled from information from *Light as an Ecological Factor,* edited by R. Bainbridge, and *Elements of Ecology,* by G. L. Clarke; Chatto and Windus Ltd., for four tables from *Farming Systems of the World,* by A. N. Duckham and G. B. Masfield; Food and Agriculture Organisation of the United Nations for a figure from *World Forest Investory,* 1965; Her Majesty's Stationery Office for two figures from *Forest Management Tables,* Booklet No. 34, 1971, by the Forestry Commission; a figure and two tables from unpublished Forestry Commission Working Plans, reproduced with the permission of the Controller of Her Majesty's Stationery Office; Her Majesty's Stationery Office and Pergamon Press Ltd., for our Figure 4.4, compiled from information from *Forest Management Tables,* Booklet No. 34, 1971, by the Forestry Commission and *The Principles of Forest Yield,* by E. Assmann, 1970; Ordnance Survey for a heavily adapted Ordnance Survey Map; Oxford University Press for Figure 4.1 from *Grassland Ecology*, by C. R. W. Spedding, © Oxford University Press, 1971; Pergamon Press Ltd., for a table from *The Principles of Forest Yield,* by E. Assmann, 1970; Royal Holloway College, University of London for a figure 'Auxin Relations to the Woody Shoot' from *Annales of Botany,* 23, by E. S. J. Hatcher, 1959; Scientific American for a figure, p. 100, from *The Nutrient Cycles of an Ecosystem,* by F. Herbert Bormann and Gene E. Likens, © October 1970 by Scientific American Inc. All rights reserved; Springer Verlag for a figure and a table from *Vegetation of the Earth,* by H. Walter.

CHAPTER 1
THE MEASUREMENT OF PLANT GROWTH

1.1 THE PHENOMENON OF GROWTH

Growth is a readily recognisable characteristic of all living organisms both plant and animal. It is particularly associated with juvenile stages of the life history of an organism; for example, the very rapid growth of newly-born babies, or equally graphic, the growth of plant seedlings when kept in a warm, moist atmosphere. It is equally expected for growth to stop when some predetermined size range is reached; a common garden cabbage of the *Brassica* family for example does not grow *ad infinitum* but stops when it has attained a specific, genetically determined shape and size. But does growth stop when a visible increase in size is no longer apparent? The answer is very definitely no. There appears to be a strict relationship between the optimum size range and optimum internal organisation. Throughout the time an organism is actively increasing its size it is also engaged in internal, and hence invisible, repair and replacement work. There comes a time when the repair work assumes such dimensions that little or no energy is available for further growth – and hence an increase in visible dimensions stops. This brief statement has been an oversimplification of events, for superimposed upon the growth curve there is also the factor of maturation of the organism. All plants and animals pass through distinct stages during their life history. This can be best shown in diagrammatic form (Fig. 1.1).

It is during the juvenile and early maturity stages that visible growth occurs. Any energy which remains after internal repair work is completed is then diverted towards

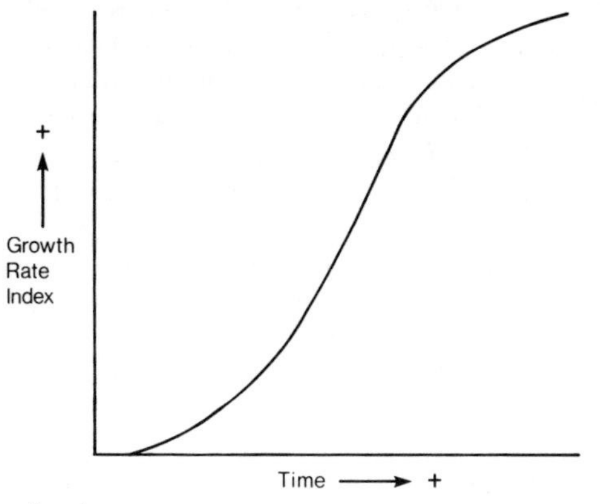

Fig. 1.1 Idealised growth rate curve

the reproduction of the species. This book is concerned entirely with plant growth and its measurement. Plant growth differs from animal growth in that it is simpler because there is no energy loss via locomotion and body heat regulation while respiration losses are also lower in plants than for animals.

The growth of plants has long been of interest to botanists. The study of plant morphology, that is the visible changes which occur during the life history of the plant, represents one of the oldest studies of botany. It has only been since the 1930s that botanists have been able to explain growth processes through improved physiological and biochemical knowledge. It has been in this latter area that remarkable advances have been made in our knowledge and understanding of plant growth.

There is one other major reason for concerning ourselves with plant growth – a more applied reason. Plants form the food source of all higher life forms on this planet. It is their ability to generate their own foodstuff from basic salts, water and sunlight in the process known as photosynthesis which makes them unique. Plant food forms the chemical energy supply of all animals, including man. No matter how scientifically advanced and technologically capable modern man may be, he is still totally reliant upon the ability of plants to fix energy and thus to grow, and to pass this growth product through food chains thereby providing a food supply for man. Plant growth thus assumes a massive significance for man. It has economic and political implications, it has influenced the social development of man, it forms the basis of massive food processing industries, provides agricultural employment and security for millions and yet in most people's minds the phenomenon of plant growth receives scant attention.

It has not always been so, however, for our ancestors were obliged to be subservient to the fluctuating whims of nature. Years of plenty alternated with times of famine. Only very slowly did organised sedentary agricultural tribes manage to attain a measure of stability in the yield of their crops. Even today agriculturalists are faced with good and bad crop yields. True, the magnitude of the fluctuations is smaller than in the past but still the unpredictable nature of plant growth and crop yield has immense implications for man. Perhaps these implications are greater now than ever before, for a proliferating world population makes ever increasing demands for food. Rarely are lack of commodities small-scale, local affairs, for shortages can now be international in their impact. Perhaps this was first realised in the 1950s when disastrous grain harvests in the USSR made the Soviet population dependent on North American wheat. World prices for grain rose sharply as a world shortage was forecast. Again and again the problem has been encountered each time with different commodities falling into short supply, sugar in 1974, potatoes in 1976, coffee, 1975–77. How far these 'shortages' are the result of complex international trade agreements between the producer and consumer nations is difficult to judge. Poor growing conditions were undoubtedly the cause of poor yield in many instances but ever-changing consumer patterns make interpretation of data difficult. The majority of the plant species used in agriculture are annuals, that is they take one year, or less, to grow from seed to harvesting time. All the cereal crops, legumes, root crops and *Brassicas* are annuals. Less common are the biennials, those which require two years to reach harvesting time. Sugar cane is an example. In a completely different category are the perennial plants; these possess the ability to grow in the year of planting, to survive an inhospitable period of the year (winter or arid phase) and to resume growth when favourable growth conditions return. This sequence can be repeated many times, often for scores of years, even hundreds of years and occasionally, as in the case of some trees, for thousands of years. At the onset of each favourable growing season additional growth is made and added to the previous growth. This may appear as a contradiction to a statement made earlier in the chapter. Perennial plants, notably woody species, i.e. trees, are capable of increasing their size indefinitely until death and decay reverse that trend.

Trees, then, appear not to have distinct maximum size ranges. Instead, they continue to grow until they can no longer be supported by their immediate environment or until man removes the tree for a timber supply. Because trees remain *in situ* for a long period of time they become useful indicators of the growth potential at any one site. The tree becomes adapted to the environment in which it grows and provided the tree is older than about 30 years, the overall growth rate can be used as an accurate indicator of the growth potential. Annual plants are far less suitable for measuring the growth potential of a site for the between-year variation combined with the between-plant variations make comparisons and the calculation of growth potential curves a dubious exercise.

The characteristic of growth can take many forms. Most obvious is the visible increase in stature of a plant – this may take the form of an increase in height, or in girth, or in the abundance of leafy tissue. Underground, and hence invisible, increase may also occur; for example, an extending root structure or an enlarged food store such as a rhizome or bulb. Other invisible growth may take the form of internal repair, replacement and modification while yet another form of growth can take the form of the creation of flowers with male and female components, hence facilitating reproduction. Thus, what was originally considered to be a straightforward concept has now appeared as a more complex entity. To understand the phenomenon of growth more fully it is necessary to explore the mechanisms of growth.

1.2 THE MECHANISMS OF PLANT DEVELOPMENT

The development of a plant from a germinating seed through to a mature adult plant requires both growth and differentiation. At the outset, growth involves an increase in the number of cells forming the organisms until eventually there exists a mass of undifferentiated protoplasm. Such a state rarely lasts for long. Not only is there a numerical increase in the number of cells but there is also a functional differentiation so that some cells become, for example, specialised root cells, others shoot cells and yet others leaf cells. Growth and differentiation are at one and the same time different processes and yet are inextricably linked to one another.

The growth of plant cells involves the uptake of relatively simple, widely available substances notably basic salts and water which are combined as a result of radiant energy to form synthesised sugars. These substances are then incorporated into cell tissue and it is this generation of new organic material from inorganic substances which forms, very basically, the essentials of growth. As more and more synthesised material is accumulated in each individual cell then so an increase in size occurs until an optimum size is attained whereupon the cell divides into two and growth begins anew. Sequences such as this can be found in simple, undifferentiated organisms such as *Chlamydomonis viridis*. Higher plants, however, do not behave in such a straightforward fashion. Not all cells in a plant are capable of division for many cells have undergone differentiation and as a consequence have become thickened from the addition of cellulose or lignin. This is best seen in trees where the bulk of the cellular material within the woody portions of the trees has been strengthened via the deposition of lignin and thus cell division is impossible over the major area of the tree. Instead, growth is confined to very specialised localities – the meristem areas, occurring at the tips of roots and shoots, in leaves, flowers and fruits. Not all meristems function in the same way. Those located in root and stem tips usually have the capability of perpetual cell division. Thus, growth is added at each of these so-called apical meristems for as long as manufactured plant food is available. In the case of perennial plants this apical meristem may remain active for many hundreds of years, the growing point advancing away from tissue of increasing age (Fig. 1.2). These meristems have been termed indeterminate meristems. They have the

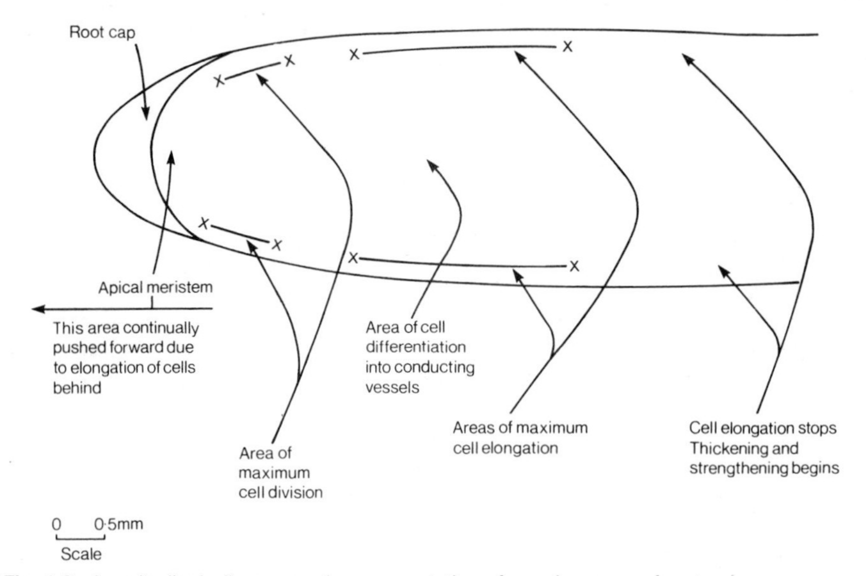

Fig. 1.2 Longitudinal, diagrammatic representation of growing zone of root

ability to branch and this gives rise to the typical, infinitely variable dendritic pattern of both the branch and root network of plants. In contrast, there exist the determinate meristems which are found in leaves, flowers and fruits. These remain active for only limited periods for when the shape and size of the organ assumes its mature shape then growth stops. Some plants, notably the very successful grasses, possess a very different type of meristem. Here, additional meristems, called intercalary meristems, are located at the internodes (the slight swellings which occur at intervals along the stems), and in the leaf sheaths. Thus growth can occur in the basal area of the plant as well as from the apical meristems. The grasses are the one group of species which can actually thrive under continuous grazing or cutting pressures, for the removal of the apical meristems only serves to stimulate the intercalary meristems.

The actual increase of size which is indicative of growth can occur either because of the increase in cell numbers via cell division or by increase of cell size. Figure 1.2 shows that there are distinct areas of the organ in which these features may be observed. Although it is by no means certain, it appears that not all the cells produced in the growing tip are meristemmatic (Wareing and Phillips, 1970). At best it appears that most of the new daughter cells are capable of only a limited number of cell divisions. Instead many of the cells begin to find themselves literally 'left behind' as the apical point continues to initiate new tissue. These cells, which are now away from the centre of active division, begin to absorb water and undergo elongation, often increasing in volume by 100 or even 150 times the original size. This process of elongation is called vacuolation; at the same time there is also a considerable increase in the amount of cell wall material present in each cell. These two mechanisms whereby growth is achieved are not absolutely confined to distinct areas. Cell division can occur in cells which have begun active vacuolation. Both cell division and vacuolation create a greater surface of cell wall area and there are two possible ways this can be achieved. First, the cell wall may simply stretch and thus become thinner. Alternatively, cell wall thickness can remain constant; if this occurs then obviously there must be considerable and rapid generation of new cell wall tissue. The second of these possibilities appears to be the adopted method. Thus growth involves an increase in size due to cell division and elongation, an increase in mass due to a greater water content and also a greater protoplasm content. In an attempt to eradicate the effect of water content on the increased weight of a plant it is customary to dry the specimen in an oven at 100°–105°C for 24 hours in order to obtain the dry weight. The dry weight indicates the amount of protoplasm which constituted the cell walls.

1.3 PHYSIOLOGICAL AND BIOCHEMICAL MECHANISMS AND THEIR CONTROL OF PLANT DEVELOPMENT

The consideration of growth has, so far, been concerned solely with the relatively straightforward and well known characteristics of cell division, vacuolation and differentiation. It is now necessary to examine how these basic growth processes are initiated.

In order that a cell may assimilate, grow, divide, vacuolate and differentiate in a precise manner and the process be replicated by countless other cells there must be a reliable coordination system built into the plant. Two very different types of coordination systems are known to exist although there may be others. First there are the 'chemical messengers', the presence of which causes specific functions. This group includes three major classes of what are generally called growth-promoting hormones; for example, the auxins, gibberellins and cytokinins. The second coordination system is the field or physical force system about which only little is known. Field forces are thought to include electrical gradients, pressure differences and gas-exchange gradients over part, or the whole, of the plant. Field-force physiology is still very much a research study and falls beyond the scope of this book. The chemical activators of growth, however, are now very well known and may be quickly reviewed. American botanists refer to these substances as growth regulators (Leopold, 1964), maintaining that the substances are not synonymous with the chemical activators found in animal tissue and to which the name 'hormone' was first given. Wareing and Phillips (1970) argue that sufficient evidence has now been amassed to show that the term 'growth hormone' is more appropriate, for, like the true animal hormones, the plant hormones are now known to be produced in exceedingly small amounts and have effects at locations distant from their place of production. A major difference, however, is that plant hormones appear capable of being multi-responsive, that is, a single hormone can produce different stimuli depending upon the type of organ upon which it encounters whereas animal hormones tend to be specific in their action.

The first recorded account of growth influencing substances can be found in the work of Charles Darwin in 1897 who related the phenomenon of phototropism to an 'influence'. The effects of phototropism will be well known to anyone who has grown a pot plant indoors on a window sill. After a period of time the growing tip of the plant will show a pronounced movement towards the source of maximum light intensity. The distortion of the shoot occurs not at the apical point but some distance below. Several decades of research were required before a chemical substance, identifiable as indoleacetic acid, was shown to control the direction of growth of the apical point. When illumination was non-uniform then the hormone had an uneven distribution, causing one side of the shoot to grow faster than the other so producing a distortion towards the light sources. The name of this growth substance produced by the growing points of all higher plants is called auxin (from the Greek, Auxein, to grow). Auxin is the fundamental hormone responsible for all growth and differentiation and can be found in both the indeterminate and determinate meristem areas. One of the continuing marvels of the growth phenomenon is that despite its obvious complexities there is almost perfect coordination of the relevant stages of growth with the passage of time; leafy tissue precedes flowers while the root structure develops in harmony with the above-ground portion. The auxin-group of hormones appear to be responsible, at least in part, for this sequence though the exact way in which the correct sequence is initiated is not understood. Hatcher (1959) showed that in spring-time apple trees showed an auxin surge approximately two weeks before visible growth occurred, by August auxin production had fallen away drastically and by November was present in almost zero amounts (Fig. 1.3).

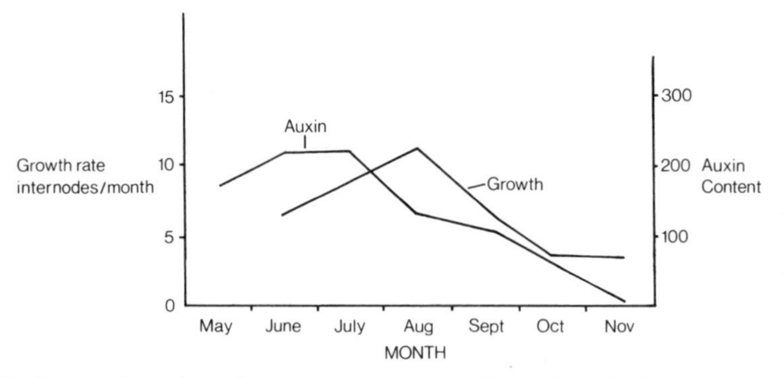

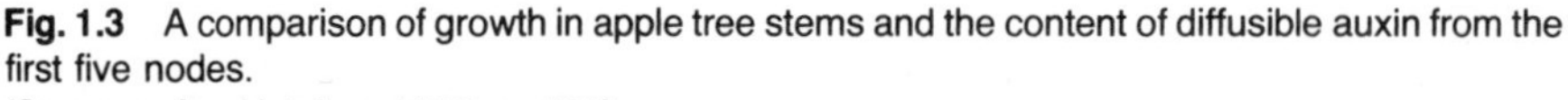

Fig. 1.3 A comparison of growth in apple tree stems and the content of diffusible auxin from the first five nodes.
(*Source:* after Hatcher, 1959, p. 409)

The precise growth mechanism prompted by auxin is simple in the extreme. Auxin causes a softening of the cell wall, so allowing an increase in cell size by osmotic uptake of water until the resistance of the cell wall is equal to the osmotic pressure. Experimentation has not yet revealed how the plant determines the quantity of auxin to produce; too little auxin would prevent growth while too much would produce over-soft cell walls with the associated risk of tissue rupture.

While a great deal of experimental evidence has accumulated since the 1930s when research into plant hormones began there are still a host of problems to unravel. The dictum of Went in 1926 that 'no auxin, no growth' has now been cast into doubt. While auxin is the outstanding known control mechanism of development in plants there is also abundant evidence to show that other substances are necessary for successful plant growth.

At about the same time that research work on auxins was revealing the significance of that substance on plant development, research in Japan was proceeding on what was to become another major chemical component of plants. Kurosawa (1926) was engaged upon the study of a disease of rice plants caused by a fungus *Gibberella fujikoroi.* The effect of this fungus was to make rice plants grow very tall, to become spindly and finally to fall over through lack of support tissue. It was not until the 1950s that the Western world learned of the now-named Gibberellin and research activity initiated into a whole group of gibberellic acids (GAs). Nine separate GAs have been identified, six from the original fungus while three more have been discovered as natural substances in higher plants. The effect of GAs on plants is dramatic; cabbages which have gibberellin added to them grow to 2 m in height while dwarf beans become climbing beans! The exact effect of the GAs on plant cells is debatable. Abundant evidence exists to suggest that the hormone causes massive new cell generation in the meristems though there are also experimental results to show that cell elongation is the result of gibberellin. Biochemists soon discovered that these hormones had a chemical structure remarkably similar to the terpinoids which are widely distributed in the resin of conifers. From a review of research work (Leopold, 1964), the gibberellins were shown to be naturally-occurring, growth-regulating substances. They are particularly connected with events such as the initiation of flowering and the ending of dormancy. These events are usually attributed to the operation of environmental variables and thus the conclusion can be reasonably made that the production of the gibberellin groups of hormones is related to specific environmental factors, most probably those of temperature and day-length.

One other promoter of the growth of plant cells must be mentioned, that of kinetin, or more accurately, a group of substances called cytokinins. As far as is known, kinetin is not found in any plant for kinetin, itself, is a synthetic substance. Substances which have similar effects to kinetin can be found in immature fuits and in particular in the fluid coconut milk. It appears that kinetin operates in conjunction with the auxins and stimulates a marked increase in cell division rate. The group as a whole has a wide-ranging series of effects on plants but its most well-known function is to ensure the breaking of dormancy in seeds and to assist in leaf growth and apical shoot development. One peculiarity of this group is their extreme immobility; application of kinetin to a plant causes localised cell division. This fact makes it almost certain that the cytokinins, as a group, are not true plant hormones.

Any study on the topic of growth soon comes upon the question 'Why don't plants grow to immense size – perhaps never even stopping their growth cycle?' Even the woody perennials appear to have a finite size, in contradiction to the statement made in the opening paragraphs, and yet this size limit may be a function of age, or of the inability of the tree to produce a sufficiently strong support system, or again may be related to nutrient circulation. It can be argued that if the growth-promoting hormones operated in field situations as they do under laboratory conditions then the whole biosphere would be tangled vegetation mess. But that is not so. There exist very definite plant inhibitors, some of which are biochemical while others are environmental.

The unlimited operation of growth-promoting substances in plants could lead to very obvious disadvantages; for example, plants which were so tall that they failed to support themselves, or overproduction of leaves or flowers. Equally, wherever plants grow in a seasonal climate, and this accounts for the majority of living space on this planet, then climatic conditions are positively harmful for plant development at certain times of the year. The growth-promoting hormones are therefore countered by growth inhibitors, the specific actions of which are unclear but seem to form an integral part of dormancy conditions in plants. They also appear to be regulated by environmental stimuli; for example, decreasing temperature and day length, increasing soil wetness. The known inhibitors have diverse molecular structures which have been taken to indicate a diversity of physiological processes over which they have some control.

The most common inhibitors appear to be controlled by climatic parameters; for example, seed and bud dormancy is the result of accumulation of an inhibitory chemical substance which occurs during the growing season and which, at the onset of autumn, has accumulated to such an extent that it causes growth to stop. Dormancy is only broken after exposure to a specific amount of cold or drought. Only after a specific number of cold days can the inhibitory substance be removed and the growth-promoting substances then begin new season growth. Thus, 'chemical clocks' triggered by environmental characteristics appear the most likely ways in which plants can regulate their growth mechanisms.

Detailed attention has been given to these chemical promoters and inhibitors of growth for although much research has been done on their modes of operation and interaction, it is certain that the biochemist will be able to achieve considerable gains in knowledge in the near future. The agriculturalist, aided by the complex plant hormones, may be soon able to eliminate weeds from crops, or to increase the size of the flower and seeds of a plant at the expense of the stem. To date the main application has been to produce dwarf plant species; for example, Amo-1618, an inhibitor with anti-gibberellin properties, has allowed much reduction of stem tissue to have been achieved but with no loss of leaf or seed production. Dwarf varieties of grain crops can thus be successfully grown at sites which had proved marginal when sown with conventional varieties. In this way knowledge of plant hormones is allowing an increase in plant efficiency and production.

1.4 ENVIRONMENTAL CONTROL OF PLANT DEVELOPMENT

The impact of environment upon vegetation growth is well known (Geiger, 1957). The term environment implies the surroundings of the plant through and from which come stimuli which can promote and retard growth. As the environmental inputs vary then so too does the vegetation; for example, compare the patterns of plant growth in southern Britain with those of the Mediterranean coastline. Many of these differences are due to man's actions – but man's actions are, in turn, influenced to a great extent by the environment. Thus in southern Britain, an evenly distributed rainfall combined with cool summers and mild winters encourage the almost year-round growth of grass. The relief factor rarely assumes a dominant dimension and slope angle, run off and erosion problems are hence only local, small scale problems. By contrast, the Mediterranean coastline is characterised by hot, sunny, dry summers and cool, damp winters. Such a climate is thus markedly seasonal with a drought period lasting from one to six months as one moves from west to east. Under such conditions plant growth can only occur during the winter period when moisture is available. In summer the plants are dormant. Mediterranean ecosystems are much more liable to destruction than is the southern British example, for in the Mediterranean, flash flooding brought about by localised thunderstorms cause extensive soil erosion. Vegetation cover is often discontinuous and appears drab compared with the green fields of southern England.

Such observations are easily obtained by a brief examination of the vegetation. Other more meaningful effects of environment will be readily understood by the gardener or farmer. The planting and havesting date of crops will vary from location to location – usually the more northerly, elevated and exposed a locality then the shorter is the growing season. Then, plant growth will vary at any one site from year to year depending upon the variations of the weather pattern. An excess of rainfall may be detrimental to one crop, say wheat, while an adjacent potato crop may benefit. Vegetation then responds in a whole variety of ways to the impact of environmental stimuli. Unfortunately, the nature of the response is difficult, perhaps impossible, to predict. Only rarely can a situation be found wherein one environmental variable changes and the remainder stay constant. Instead, a change in one variable often leads to a chain reaction with other variables changing in response to the first. For example, a cloud cover will lead to a reduction in radiation levels followed by a reduced temperature which produces a change in the relative humidity of the atmosphere. It is unlikely that a plant displays any uniformity of response to such changes. Members of the same plant species, even of the same variety, will respond differently, often depending on the development stage of the plant.

Even more fundamental than the applied effects of environment upon vegetation as recognised by agriculturalists, horticulturists and foresters is the invisible impact of environment upon plants. Environmental stimuli provide the means whereby the entire metabolism of a plant is synchronised so that each life history unfolds in a correct sequence. The biochemical mechanisms reviewed in the previous section are thought likely to be under the control of environmental factors. Thus, environmental cues are probably responsible for the time of day that a flower opens and closes, for the onset and breaking of dormancy and for the orientation of leaves to the sun's rays.

It is useful to review five macro-environmental variables *viz*: solar radiation (and its dependent derivatives of light and temperature), precipitation and soil. The interplay of these variables determines the general plant growth pattern.

Solar radiation is constantly received upon this planet though at any one site there will be distinct periods of radiation input. At night time the radiation levels will be very low, indeed, they may even be reversed with a net loss of radiation from the ground out to space. The radiation input provides energy which is incorporated into the plant protoplasm. The radiation will be of varying wave-length, (Fig. 1.4) and as a general rule the plant cells are capable of selective absorption of the wave-lengths, in particular of the ultra-violet and visible light bands.

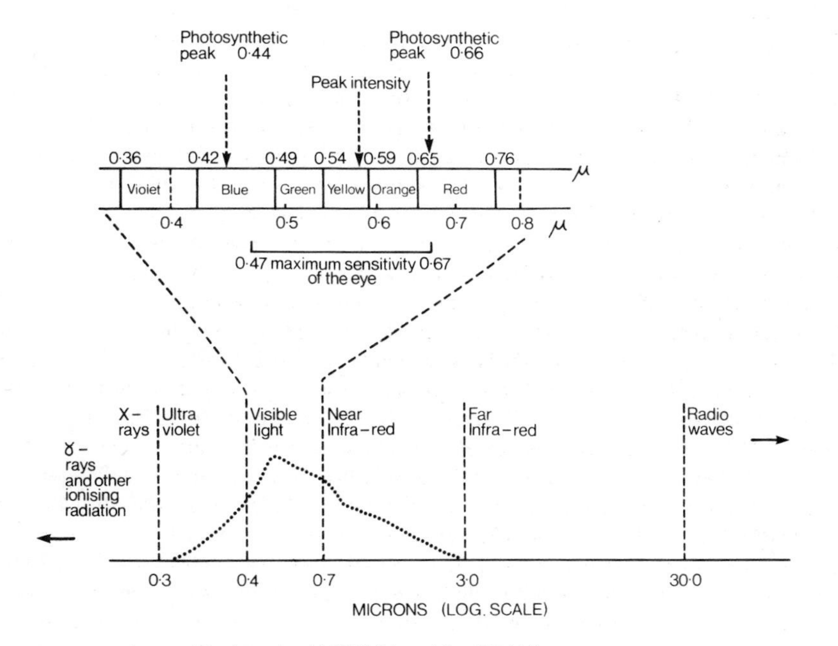

Fig. 1.4 Spectral distribution of radiation as a function of wave-lengths
(*Source:* after Clarke, G. L., 1957, p. 185, and Blackwell, M. J., 1966, p. 17)

The protoplasm components of the cell walls possess electron-resonance properties which allow absorption of specific wave-lengths, e.g. surface water plankton are responsive to red and blue light because they contain large quantities of chlorophyll 'a' and phycoerythrin. As wave-lengths become shorter, the selectivity becomes substantially reduced and X-rays are almost uniformally absorbed by all portions of the plants. These shorter wave-lengths are also the high-energy wave-lengths and consequently become damaging to plant tissue with the nucleic acids and auxins being extremely susceptible. Gunkel (1957) has reviewed the work that has been made upon the effects of radiation on plant growth. Radiation can produce a wide range of effects depending upon the wave-length, duration of exposure, repeated exposure, and stage of growth of the irradiated plant. Budless shoots, tumours, callus formation and destruction of the meristems are common effects of radiation. Fortunately, under normal environmental conditions, damage to plant tissue from the short wave energy is rare because the atmosphere provides a filtering mechanism which removes or reduces the harmful X-rays and gamma rays. Man's actions must be briefly mentioned in this context. The increase in use of nuclear energy for both peaceful and warfare uses has led conservationist groups to argue that a nuclear accident would lead to massive fallout (radiation). Such an event would inevitably destroy or deform vast amounts of plant and animal tissue, though Mellanby (1967) has shown that not all radiation damage must be considered harmful. It is probable that radiation has exerted a very strong influence upon the evolutionary development of organisms through the stimulus of genetic modification.

Much of the radiant energy not only has a capacity to heat the biosphere but also to illuminate it. While the energy input is principally in the wavebands 0·300 to 2·000 microns by far the greatest amount lies in the 0·36 to 0·76 microns band – the so-called 'visible light band'. The remaining wave-lengths are substantially reduced by atmospheric filtering. The visible light band is unique in that it does not usually prove harmful to protoplasm whereas, as already shown, other wave-lengths particularly below 0·36 microns are extremely hazardous. It is probably no coincidence that both plants and animals have evolved to utilise these abundant wave-lengths. Plants, for example, have evolved so that the individual cells contain pigments which are responsive to individual wave-lengths. This means, in effect, that for plants to grow they must be triggered by the correct light inputs. It is now known that all the main stages in the life history of plants are light dependent, e.g. photosynthesis, tropisms, photoperodism and even pigmentation are all controlled by receipt of specific wave-lengths. Photosynthesis, for example, is most efficient at 0·44 and 0·66 microns (Fig. 1.4). The overall control of light on plant development from the manufacture of food through to the control of the quality and quantity of growth has established light as the most significant of environmental variables.

Incoming solar energy can also be significant in that besides its properties of radiation and illumination it is also a source of heat. When the electromagnetic waves strike an object, be it a plant leaf or a concrete wall, the energy becomes disorganised and heat is generated. Considering the vast range of temperatures which can exist, plants show a very narrow tolerance range for temperature, usually within the 0°–50°C range. This 50C° range is just about 0·5 per cent of the temperature range over which atoms can exist. The range is due to the freezing point of water at 0°C while at about 50°C protein substances begin to disintegrate which leads to a breakdown of cell entropy.

Plant surfaces, notably the leaves, are subjected to a severe heat-exchange problem. Russel (1961) has calculated that in southern England the cumulative annual insolation is approximately 76,000 cal/cm²/yr. That is equivalent to 8·5 Mkw hours of electrical heat/ha of ground or the burning of 988 tonnes of coal/ha. Leaves utilise heat energy in one of four ways: it may be transmitted by absorption, it can be lost by conduction, it can be re-radiated directly to space or it can be utilised in the process of transpiration. One of the most obvious relationships between temperature and plants is that the rate of growth is strongly influenced by temperature. At a temperature of about 6°C growth process begin and meristem activity becomes measurable. As the temperature increases then so growth rate increases – providing that other environmental requirements are available, notably moisture. When ambient temperature reaches about 35°C growth rate is at a maximum, any further temperature increase generally produces a growth rate decline because of the accelerated disintegration of cell proteins. Note that no definite temperatures can be quoted which mark the beginning and end of the growth phase. Temperature values of about 6°C and 35°C appear to be the significant controlling values for a great many plant species (Taylor, 1958).

Extreme temperatures, that is below 0°C or above 50°C, can produce severe thermal damage though species show differing abilities to resist such extremes. Many domesticated plants – notably some *Brassicas,* and many of the *Coniferales,* the coniferous trees – exhibit an ability to adapt to low temperatures. In the autumn, when the first frosts check and then kill many species such as the annual weeds, the resistant species mentioned above survive relatively undamaged though growth does stop. Work by Day and Pearce (1936) and Forestry Commission (1946) has shown that frost resistance is associated with a large number of different cell constituents and conditions but that in many cases the association is a loose one and may vary from species to species and upon the time of year. Damage from high ambient temperatures appears less common than from low temperatures probably because air temperature rarely attains, or is maintained, at values of 35°C or above for more than 12 hours at a time whereas long-lasting low temperatures are commonplace in continental interiors during the winter time. It should be noted, however, that leaf surface temperatures may record an additional 15C° above ambient temperature provided the leaf is in receipt of strong sunlight. Leaf temperatures of 60°C have been recorded under sunny, still air conditions (Geiger, 1957).

There is little large-scale intervention that man can make upon temperature alleviation. Frost prevention measures in orchards and vineyards are commonplace while glasshouse cultivation is perhaps the ultimate in temperature control. All temperature control methods are expensive in terms of energy, material and labour consumption and are confined to high-value crops. The bulk of the plant species are allowed to respond to temperature changes in the best way they can – by inbuilt genetic response. The Forestry Commission in Britain have clearly shown the importance of seed provenance in respect of resistance to low temperatures. Sitka spruce (*Picea sitchensis*) seed from Queen Charlotte Isle, BC, has proved best to meet the British Isles environment and in particular the temperature hardiness appears particularly suited to British conditions.

Solar radiation has been already credited as being the most significant of environmental variables (see p. 9). Without light, plant growth would be impossible. But one other environmental input also assumes a significance very nearly as great as light – that of water. Water forms between 80 and 90 per cent of most living cells, much of it being used to provide turgor pressure within the plant which in effect allows the plant to stand erect. Perhaps of greater significance are the water molecules which are incorporated into the protein and nucleic acids which exist in the individual cells. Water is also an efficient solvent, dissolving all the minerals contained in the soil and conveying them into and through the plant tissues. Water is a reagent in biological reactions in that it is a substrate for photosynthesis and forms a major product of respiration.

Water and/or water vapour plays a critical role in the balancing forces of the biosphere. The constant transfer of water from the oceans to the atmosphere and eventually back to the oceans or land in the form of precipitation assists in equalising temperature and moisture gradients. Water vapour in the atmosphere is vitally important in absorbing the potentially hazardous short and long wave radiation (Fig. 1.4). The oceans can absorb considerable amounts of heat energy without changing much in temperature. The property of water to absorb heat energy and yet for the temperature to change little makes water a supremely suitable liquid to fill cell spaces. It allows plants to exist in a state of almost constant, or at worst, slow changing temperature conditions.

Plants have evolved over an immense amount of time so that water enters the plant via the root hairs and travels through the plant in response to an energy gradient, and eventually is expelled from the plant most commonly as water vapour via the minute stomatal openings on leaf surfaces. From the purely ecological viewpoint the movement of water through the plant is unimportant – but the entry and loss of water are of great concern for these are processes that are strongly influenced by the environment.

Moisture is usually provided to plants by natural precipitation via the soil. Most vegetation types require an evenly distributed moisture supply during the growing season though some vegetation has become adapted to survive seasonally dry conditions. The demand by a plant for moisture is never constant. In the active growing season demand for water is great while in a period of dormancy, caused perhaps by insufficient illumination or low temperatures, water demand is also reduced as growth processes will cease. Actively growing plants can transpire large quantities of water each day; the humble cabbage consumes more than a litre of water per day while an oak tree may transpire 570 litres (Daubenmire, 1974). A vegetated surface can lose a far greater volume of water than an adjacent bare soil surface and work by Penman (1948) has shown that considerable areas of even a maritime country such as Britain often experience a severe shortage of soil water during the late summer stages of the growing season. This soil water shortage can cause a temporary cessation of growth and in extreme cases will cause wilting of plants and even death. A common feature of larch plantations (*Larix* spp.) growing on steep valley sides in Wales during late summer is the drooping of the leading shoot. This is due to a substantial soil water deficit occurring on free draining valley sides. It is usually short lasting and causes no permanent damage to the tree though it is likely to cause a temporary dip in the growth curve.

The availability of water to plants is one environmental variable which can be relatively easily controlled by man. Irrigation methods can be used to add water in times of drought while a variety of drainage techniques exist to remove water where it occurs to an excess. Initial costs for both drainage and irrigation can be high though the improvements in yield can be dramatic, irrigation providing a 200 or even 300 per cent increase in yield of crops such as peas, potatoes, salad vegetables and soft fruits. Thorough drainage permits the development of an adequate root structure and this is fundamental to healthy growth. This is well illustrated in the reafforestation schemes of Britain. It is now the policy of the Forestry Commission to plough and drain all land prior to planting so that the young tree transplants can become quickly established. Without ploughing and draining the saplings would be slow growing and be prone to wind-blow because of an inadequate root structure in a water-logged soil (Forestry Commission, 1963).

It can be seen, therefore, that the presence of water, in the right quantity and at the right time, is fundamental to successful plant growth. The water utilised by plants is usually held in the soil in which the plant is rooted. The properties of soil become critical environmental variables and as such can be briefly examined as the fifth and final environmental control on plant growth.

Soil is an amalgam of weathered rock fragments, decaying plant and animal debris, collectively termed humus material as well as water and air. The variations in the possible combinations with which these components can mix are innumerable and thus soils are infinitely variable in their detailed characteristics. Despite this variation, pedologists have produced a number of soil classification systems. The most acceptable, and now most widely used system, is that derived by the US Soil Survey Staff (1960) entitled *Soil Classification – A Comprehensive System – 7th Approximation*. This system is based entirely on the properties of the soils as they occur in the field at the present day. Emphasis is, therefore, very much on contemporary soil properties and not on the evolutionary processes of soil formation. Soil properties are measurable characteristics and it is now possible to define a soil in great detail. Vast areas of soil show little resemblance to their original condition for man, almost from his earliest days, has attempted to cultivate the ground in order to grow his crops. The method of cultivation has changed somewhat from the antler-pick of Neolithic man to contemporary large-wheeled tractors and sophisticated implements. The addition of inorganic fertilizers has allowed soils even of impoverished nutrient status to be used for agriculture. Modern soil management techniques, along with improved knowledge of plant nutrient requirements, has revolutionised all forms of plant management systems. The advances have been greatest in the wealthy, industrialised nations where education standards permit the forester or farmer to understand the results of modern research. The same is not true for the developing nations where poverty and ignorance depress vegetation productivity levels. Even the most advanced land management systems are ultimately limited by the capacity of a soil to release the necessary plant food minerals. Often a soil may possess an abundance of suitable plant foods but they may exist in an unavailable form. It is the task of the soil water (and the pH status of the soil) to release the minerals for subsequent use by plant rootlets.

The impact of the physico-chemical and environmental factors upon the rate of growth of plants can be profound; for example, the essential nature of soil trace elements to permit successful growth, or the hazardous effect on growth of wind-borne sea-salt in maritime areas are but two examples from a vast milieu. There are, however, situations in which physiological, biochemical and environmental variables appear less important in controlling growth rate and instead the relationships between adjacent plants appear to have an effect upon growth rate. These relationships are generally considered to be the result of density-dependent controls based upon biotic interactions between plants. The measurement of these controls forms part of the complex phytosociological studies undertaken by ecologists and biogeographers (Kershaw, 1964; Dickinson *et al.*, 1971). It is often extremely difficult to distinguish where the density-dependent controls give way to density-independent controls; when plants are growing in close proximity there may be, for example, a density-dependent negative feedback operating to restrict seed production. This serves to prevent an increase in plant population while there may also be density-independent negative feedback in the form of nutrient depletion at root level. It is virtually impossible to be categorical about which negative feedback process is dominant at any one place – the balance of the situation may change with time.

In truly natural ecosystems a whole host of factors operate to ensure that density-dependency conditions are minimised. Species diversity is perhaps the greatest single factor which can achieve this, for different species will place different requirements upon the habitat thereby reducing the risk of critical-component exhaustion in the environment. Thus, in rain forest, a maximum species variety of up to 100 different species/ha can be recorded (Klinge, 1975) and this has probably been achieved partly because of the inherent productivity of rain forest environments and partly because maximum species variation spreads the demand for specific environmental requirements. At the opposite extreme, a man-maintained moorland typified by the Yorkshire Moors, England, will be composed of extensive areas of a few species of plants, *Calluna vulgaris, Erica tetralix, Mollinea caerula,* and perhaps *Deschampsia* spp. and sedges depending upon local soil drainage conditions. Such a monotonous flora is visually unattractive but is also ecologically disastrous. The species have common feeding requirements, have similar rooting patterns and thus produce similar demands upon the habitat. Under such conditions density-dependent factors begin to operate and reduce the rate of growth.

Competition between species of the same variety (intraspecific competition) and between different varieties (interspecific competition) represents another variation upon the density-dependent theme. Again competition for key variables appears to be the critical factor in determining the decline in the viability of a plant.

It can be seen that plant growth results from a complex amalgam of physiological, biochemical, environmental and phytosociological factors. The first two are predominantly internal to the plant–soil association, the latter two being external to the plant and operate via the pedological, biotic, climatological and anthropogenic inputs. Because of the great number of possible combinations of variables it becomes difficult, indeed impossible, to predict the result upon plant growth patterns. Even the highly organised agricultural systems of Britain are vulnerable to unfavourable combinations of these growth-controlling factors. Although much has been done to understand the ways in which individual factors work their interaction with plants remain unpredictable.

1.5 MEASUREMENT OF GROWTH

The remainder of this introductory chapter examines the main indices of growth, the alternative methodological approaches, their ease of use and their usefulness in monitoring vegetation productivity and production.

Plant productivity is the measure of energy accumulation over a specific time period and given area. Thus productivity can be expressed as the number of calories produced on a square metre of vegetation per annum, or on a larger scale, tonnes of yield per hectare per growing season. Ecological productivity is specifically the total production of an ecosystem. That includes plant, animal and decomposer productivity. The measurement of productivity even from a relatively small ecosystem such as a freshwater pond is a massive undertaking and has been successfully achieved by only a few workers (Teal, 1957; Slobodkin, 1959). Instead, research has been confined to the productivity of specific species in an effort to develop methodology and to aid understanding of productivity processes. The measurement of plant productivity is, in some ways, easier and in others more difficult than the measurement of animal productivity. For example, plants are static objects and hence do not have an energy loss via locomotion processes. Similarly they do not have an energy loss via heat generation as do the warm blooded mammals. Their respiration rates can be easily monitored. In these respects they are simpler than animal populations. However, plants do present problems because of the fact that they are rooted in the soil. It is relatively easy to weigh, at repeated intervals, an animal and thus record an increase or decrease in body weight. Such a simple method is rarely possible with plants for removal of the plant from the soil would severely check, or even kill the plant thereby precluding any more work on that individual. Annual plants can be harvested at the end of their growing season and weighed. Despite the fact that all the growth could have occurred in just a few months, or even weeks in the case of some rapidly growing weeds such as chickweed (*Stellaria media*), the amount harvested is usually considered to be the annual gain. Again, where annual plants exist in profusion it is perfectly acceptable to remove a sample at intervals throughout the growing season in an effort to identify the period of maximum productivity. Such a technique is not possible for large perennial plants. On a practical point, no laboratory is capable of weighing fully grown trees while the transporting of trees from the field to the laboratory without damaging any part of the tree would be impossible. To overcome this problem, ecologists resort to the AMOUNT of accumulated organic matter instead of the RATE of accumulation. The term production, as opposed to productivity, is given to the amount of organic matter. The production figure is often that associated with managed ecosystems, notably forestry and agriculture, while productivity figures can be quoted for small natural ecosystems. Production figures usually relate to a specific time interval, either a single growing season as in the case of food crops, or a number of years which represents the economic life of the crop, as for plantation agriculture or tree growth.

Productivity, as already stated, is the rate of accumulation of plant material. Production on the other hand can have a variety of meanings. To the ecologist the production figure will represent the total quantity of organic matter in a given area. This will be directly related to the biomass, the total standing crop containing plants, animals and microbes. To the farmer, however, the production figures will be the saleable, or usable quantity of crop produced by the end of the growing season or life cycle of a particular plant or animal. Thus yield of potatoes in tonnes per hectare will, to the farmer, represent the production figure while the forester will recognise production as the yield of saleable timber in cubic metres per hectare. Both farmer and forester will have ignored the leafy material, small branches and root systems – to them they represent the 'waste' production.

There is a major difference, then, between the terms productivity and production, while the latter term can be variously interpreted depending upon the user. Examples will be given later in this book of applications of the various uses. The simplest index which can be used to measure productivity and production is that of yield. This may be measured by weight. Thus a farmer can weigh the bags of potatoes before leaving the field on the trailer or a forester can measure the volume of saleable timber before it leaves his timber yard for the sawmills. Such a measure is, inevitably, crude for the instrument of measurement will invariably be an easy to use method, perhaps a spring balance or a linen tape – both notoriously inaccurate instruments. Added to this must be operator error – again, often considerable because of the difficult working conditions or the unskilled labour.

A different measure of production can be obtained from the calculation of the biomass. Once again it can be expressed as a volume or weight measure. Great care has to be taken over standardisation and repetition of the measurement process. Consequently, a plant is rarely weighed in its fresh condition for the fresh weight will contain a large and variable proportion of water (see p. 5). Specimens are, therefore, dried before weighing. Biomass is, theoretically, the total organic matter at a site.

Assuming that animal and decomposer groups are to be ignored one is still left with the problem of assessing the weight of root systems. To dig out, intact, a root system is exceedingly difficult and hard work. Roots are, therefore, sometimes ignored and the above-ground portion only is weighed. Once a biomass figure has been attained it is vital that any qualifications must be clearly stated before using that data for comparison with other biomass values. It is all too easy to compare unlike biomass data and consequently to draw incorrect conclusions.

The measurement of productivity is far more complex and Phillipson (1970) has reviewed the difficulties. Fundamental to the calculation of productivity is the assessment of the photosynthetic efficiency of a plant. This is the efficiency with which a plant can convert solar energy into chemical energy. Photosynthetic efficiency can be calculated in a variety of ways each giving varying results. For example, solar energy is usually interpreted as the total amount of visible light which stimulates photosynthesis in plants. The chemical energy produced as a result of photosynthesis is most usefully interpreted as net production, this being gross production minus losses brought about by respiration and internal repair work. Photosynthetic efficiency, therefore, can be expressed as a ratio of net production over total amount of light in the wavelength 0·36 to 0·76 microns which reaches the plant. This light energy forms the vital activating agent in the process of photosynthesis as displayed in the formula

$$6CO_2 + 6H_2O + \text{light energy} \rightarrow C_6H_{12}O_6 + 6O_2$$

The actual measurement of photosynthetic efficiency (the PE value) is exceedingly complex and the most useful method for plant production research involves the measurement of either oxygen or carbon dioxide production. This method is of particular value for use on grassland where proportionately large amounts of production can be lost to grazing animals (see section 6.2). In non-agricultural ecosystems it is possible to avoid the measurement of PE values and instead work with annual biomass. It is possible to compare PE and biomass values by applying appropriate conversion factors.

The measurement of productivity is thus a complex procedure and the end result is by no means free from possible error. It is necessary to provide as many cross checks as possible on the accuracy of productivity data. Some of these methods are examined in the remainder of this chapter. All are of particular relevance to the larger members of the plant kingdom and in particular to perennial species which are so difficult to assess via traditional biomass methodology.

One of the simplest alternative methods is a repeated measure of height of a plant over time. When this can be done over a sufficiently long time, height can be plotted against time to produce a graph showing rate of growth. Under some circumstances it is possible to construct standard growth rate tables based upon a large number of sample plots. Thus, repeated testing of specified crops by commercial seed companies has enabled average growth rates to be set for individual species and against which farmers can judge the performance of their own crops.

An alternative to rate of growth can be the incremental growth value. This is particularly useful when long-living plants are being assessed; for example trees, which add new wood to that of previous years. The increase in the girth or diameter of a tree is a good indicator to the forester of the speed of growth or production of new wood. Incremental increase can be easily measured by an incremental tree borer which can extract a thin core of wood from the diameter of the tree. The distance between the annual rings can then be measured and the rate of growth related to age and/or to a host of possible controlling factors. In the forest the controlling factor might include the impact of exceedingly harsh or favourable climatic spells, of insect attack or disease or of management practices.

A very interesting approach to the measurement of forest productivity has been made by the British Forestry Commission. The mean height of the 40 largest, similar aged trees is calculated to give the 'top-height' value and this top-height can be related to an index value of productivity, termed the 'yield class'. This is possible in British forests only because the bulk of the woodland is composed of even-aged, single species forest stands which have been subjected to a specific management policy. The practical usefulness of this method of calculating productivity will be examined in detail in Chapter 4.

The whole concept of plant growth and its measurement is, therefore, composed of a large number of components, some of which are now well documented, such as the effects of auxins, others are measurable but their effects on plant growth are unpredictable, such as the environmental group of components while the third group, the phytosociological components, are difficult to measure and their significance difficult to assess. One can reasonably assume that growth in plants occurs for a number of related reasons and that for man's well-being it is necessary for him to exploit considerable quantities of this growth as a food supply, or as basic raw materials in industry, for medicinal extracts and for the construction of homes, schools and other buildings.

The remainder of the book will show how the rate of growth of plants can be measured and will investigate the ways in which man can best utilise the products of plant growth. It appears certain that man will not be able to reverse his evolutionary trend as an exploiter of ecosystems and therefore it is the task of the environmental scientist to tell the industrialist, economist and politician to what extent individual ecosystems can be 'farmed'. It is for this reason that a measure of plant productivity becomes of utmost importance.

CHAPTER 2
WORLD PATTERNS OF NATURAL VEGETATION PRODUCTION

2.1 INTRODUCTION

The global distribution of vegetation types is the result of many processes most of which have been in operation for a very long time though there are also one or two relatively newer processes to which consideration must also be given. There is little doubt that the long-term impact of climatic and soil properties have had a marked effect upon the evolution of our present-day flora. Palaeontologists can show how the fortunes of whole groups of species ascended and declined over the range of geologic time from Cambrian to present times – a time span of about 500 million years. Our present-day vegetation is composed primarily of recently evolved Angiosperms which emerged sometime during the Cretaceous era some 100 million years ago. There are, however, a small number of very much older Gymnosperms (mostly *Coniferales*), and a variety of *Bryophytes, Thallophytes* and *Pteridophytes*. The evolutionary trends which have occurred in the plant kingdom have probably been due to a combination of physico-chemical and genetic relationships which have interacted with the environmental inputs outlined in Chapter 1.

The present-day vegetation distribution is, therefore, merely the end result of the operation of these long acting processes. However, over the last 500 000 years or so there have appeared additional factors, the most significant of which has been the appearance of *Homo sapiens,* man, as the dominant organism, capable of deliberately changing the vegetation pattern by deforestation, burning, ploughing, grazing and fertilizing. To this formidable list must also be added the newest methods of all – that of chemical and genetic modifications. These take the form of chemical herbicides, fungicides and insecticides while the plant geneticist can, by selective breeding, isolate desired plant characteristics at the expense of unwanted features and by so doing can thus alter the distribution pattern of plants.

Of all the environmental processes the climatic inputs emerge as undoubted master factors in the determination of global vegetation distribution, in particular the inputs of heat, light and moisture exert a direct and profound influence upon soil and vegetation. The relationships which exist between the soil and vegetation are so closely interrelated that they may be considered together almost as an entity. The vast majority of plants are rooted in the soil from which they gain both mechanical support and a supply of nutrients and water. In return, the plant when it dies, releases the chemical elements contained in its protoplasm back to the soil so establishing a cycling pattern of nutrients. The essential nature of soil nutrients is a well-known requirement for healthy plant growth (Russell, 1957). Less well known is the relationship between plants and moisture. Terrestrial plants which form the bulk of the commercially valuable plants are exposed to extreme moisture stress. Water is the medium in which nutrients are conveyed via the cation exchange process from the soil moisture films across the cell walls of the root hairs and up to the leaves where photosynthesis occurs. Here, water molecules become involved in another vital process, that of respiration which involves the release of water vapour from the leaf via stomatal openings to the atmosphere. Water also is responsible for the rigidity of plants – without internal water pressure plants would be limp, prostrate objects. Availability of water is therefore of paramount importance for successful plant growth. Ideally, water should be available whenever temperature is suitable for growth to occur but unfortunately this does not always happen. Because of the existence of complex global climatic zonation it is possible to find almost every combination, favourable and unfavourable, of temperature and moisture availability. In response to such diversity plants show a variety of adaptations ranging from continuous growth through to complete encystment in order to survive exceedingly unfavourable conditions; for example the Rose of Jericho (*Anastatica hierochuntica*).

Despite the seemingly chaotic variation in temperature and moisture inputs across the planet there is an overall, zonal climatic pattern in which temperature decreases from the equator to the poles. This pattern is mirrored by seasonality which increases towards the poles. Unfortunately, no such generalisation may be made for precipitation although rainfall patterns are related to atmospheric pressure belts which in turn are associated with the migration of the sun between the two Tropics. The pattern for light intensity also shows a complex distribution. While it is strongly related to seasonality, that is long days are characterised by an abundance of illumination, there are two other factors which negate this trend. In regions within 5° N. and S. of the equator cloud cover is usually well developed and thus it filters out much direct solar radiation. Not until latitudes 23½° N. and S. are reached do we find descending, stable air with a minimal cloud cover and a maximum solar radiation input. The second complicating factor is that of the decreasing angle of incidence at which the sun's rays strike middle and higher degrees of latitude. The increasing day length which occurs during the summer solstice in the northern hemisphere is beneficial only up to about 56° latitude N. Beyond, the increasing obliqueness of the sun's rays plus the thicker layer of atmosphere through which they travel and the greater spread of the rays negates the increase in day-length.

World vegetation patterns bear a strong relationship to the main climatic zones (Fig. 2.1). As a result of the interplay between temperature, precipitation, seasonality and illumination levels nine major climatic zones can be briefly reviewed as being relevant to world vegetation patterns.

1. Surrounding the equator and extending discontinuously 10° N. and S. thereof is the humid tropical zone. Here, mean daily temperature approximates 25°–27°C and diurnal variation exceeds annual variation. Precipitation totals vary greatly from 500 mm to 4 000 mm p.a. The outstanding feature of this zone is its lack of seasonality. The extremely low or high rainfall totals are the result of locational factors, e.g. rain shadow or exposed location under the influence of monsoonal type winds. As a result of the climatic constancy the vegetation usually displays a marked lack of seasonal variation.
2. From 10° N. and S. extending towards the tropics there occurs a semi-arid tropical zone in which seasonality becomes increasingly pronounced the further one moves from the equator. This is due to the migration of the overhead sun; rainfall and temperature reaches a maximum when the sun is at zenith while a cool, dry season prevails when the sun moves into the opposite hemisphere.

3. An arid tropical zone exists around 23½° N. and S. of the equator, extending in parts to 30° N. and S. Here may be found the true hot deserts, the Sahara, Kalahari and Great Australian Deserts. They represent regions of descending dry, stable air, the return flow of the Hadley cell which began as the vast upwelling of hot, moist air at the equator. This zone experiences extreme radiation and illumination levels, exceedingly high day time temperatures with values up to 50°C, while at night 0° is not uncommon. Rainfall comes as short, widely scattered storms. As a result of these climatic characteristics this zone has proved one of the most rigorous for the survival of vegetation.
4. Beyond the sub-tropics can be found discontinuous, transitional semi-arid, or

Fig. 2.1 Main thermal zones and bioclimatic regions
(*Source:* redrawn from Bazilivich *et al.,* 1971, p. 303.) (Reproduced with permission of the American Geographical Society)

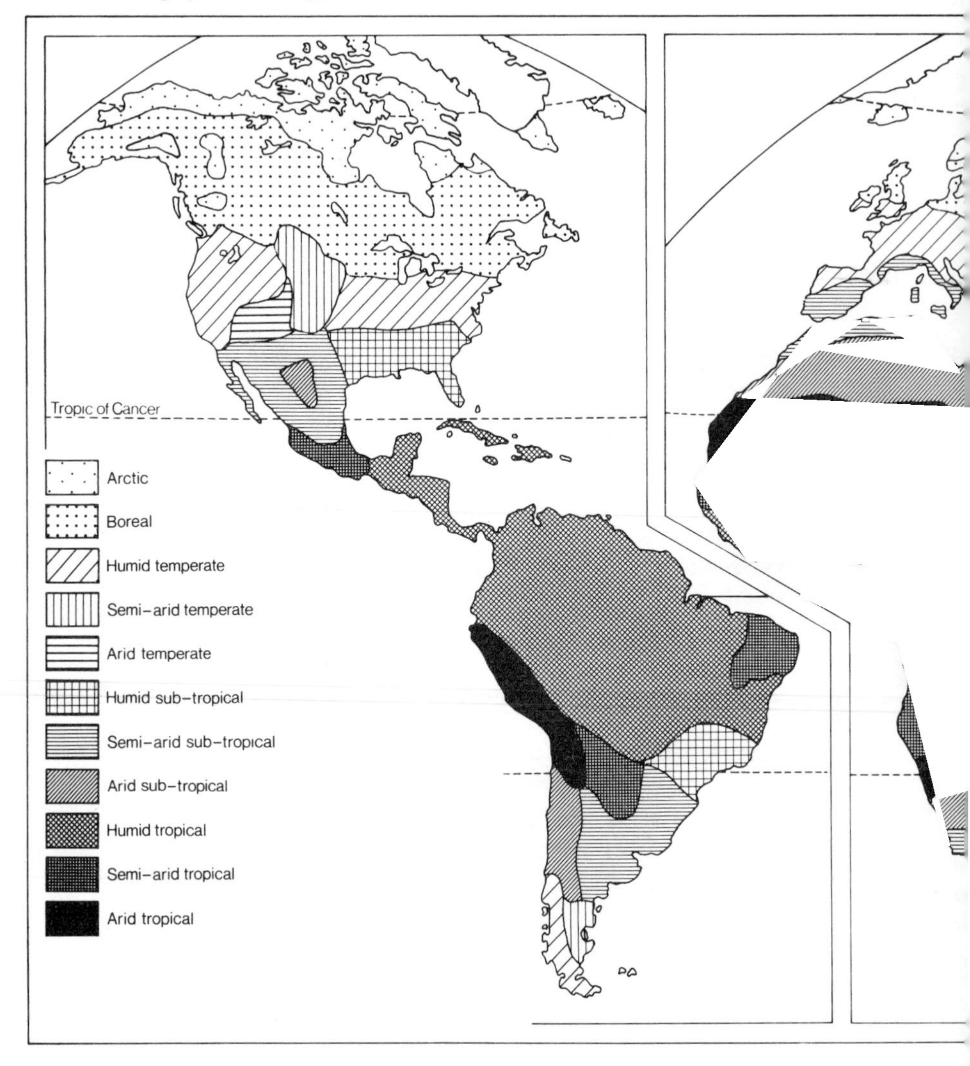

occasionally truly arid, sub-tropical areas in which the high pressure, dry air zone of the tropics predominates in summer while in winter low pressure air masses bring rain and lowered temperatures. These areas are usually described as having a Mediterranean climate though in effect there can be considerable variation depending mainly upon the distance from the sea.

5. At about latitudes 50° N. and 45° S. the impact of the dry, descending tropical air is almost totally replaced by the effects of the polar front and the consequent production of low-pressure cells. There is considerable variation once again due to detailed locational factors. It is possible to subdivide this climatic type.

5a. The humid temperate, oceanic locations. The maritime influences bestow an

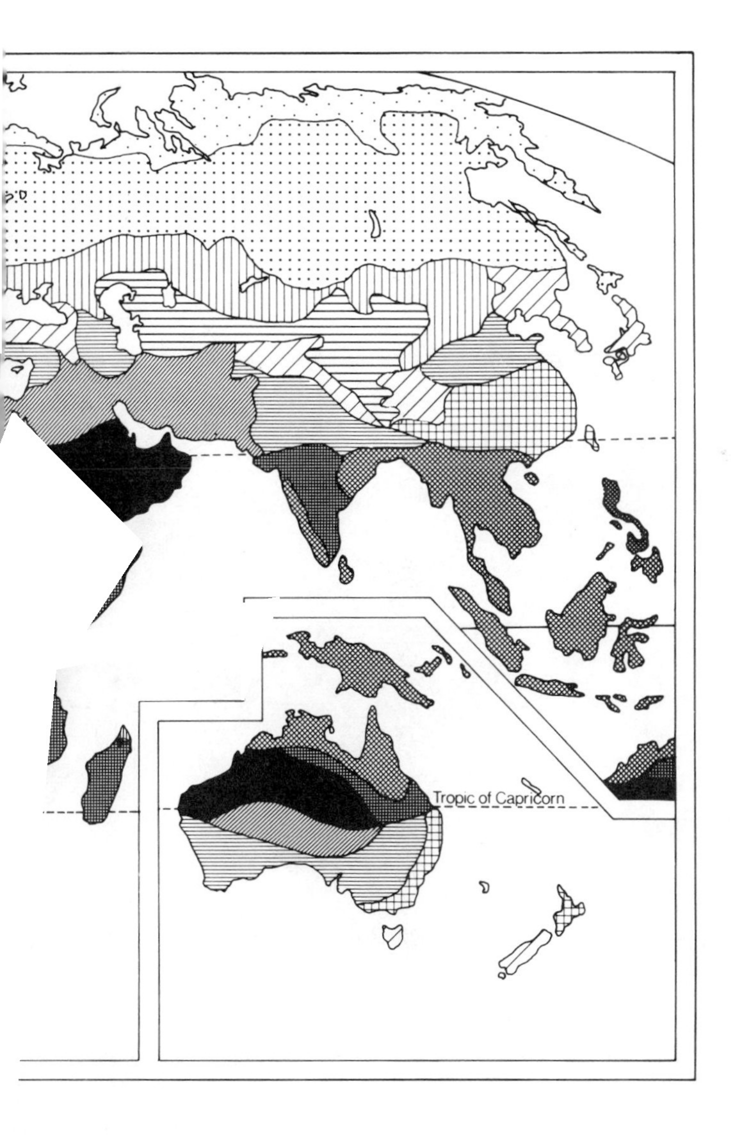

annual temperature regime with only slight variation, e.g. Scilly Isles 8°C annual variation. The mild winters, cool summers and year-round rainfall might be considered to be all beneficial for vegetation growth. These favourable conditions are, however, negated by strong, salt-laden winds which stunt, trim, burn and defoliate any tall growing vegetation and, for example, lower the tree line to only 100 m or so above sea level in western Ireland.

5b. Semi-arid temperate locations. Away from the direct influence of the sea the tree line rises rapidly as the exposure factor is reduced. Also reduced is the ameliorating effect of the ocean on temperatures and the annual range now increases, e.g. 27°C at Kiev. Precipitation comes mainly as snow in winter and from thunderstorms in summer. A greater proportion of the annual precipitation occurs in summer.

6. Extreme continental locations experience even more pronounced arid temperate

Fig. 2.2 Generalised map of world soil types

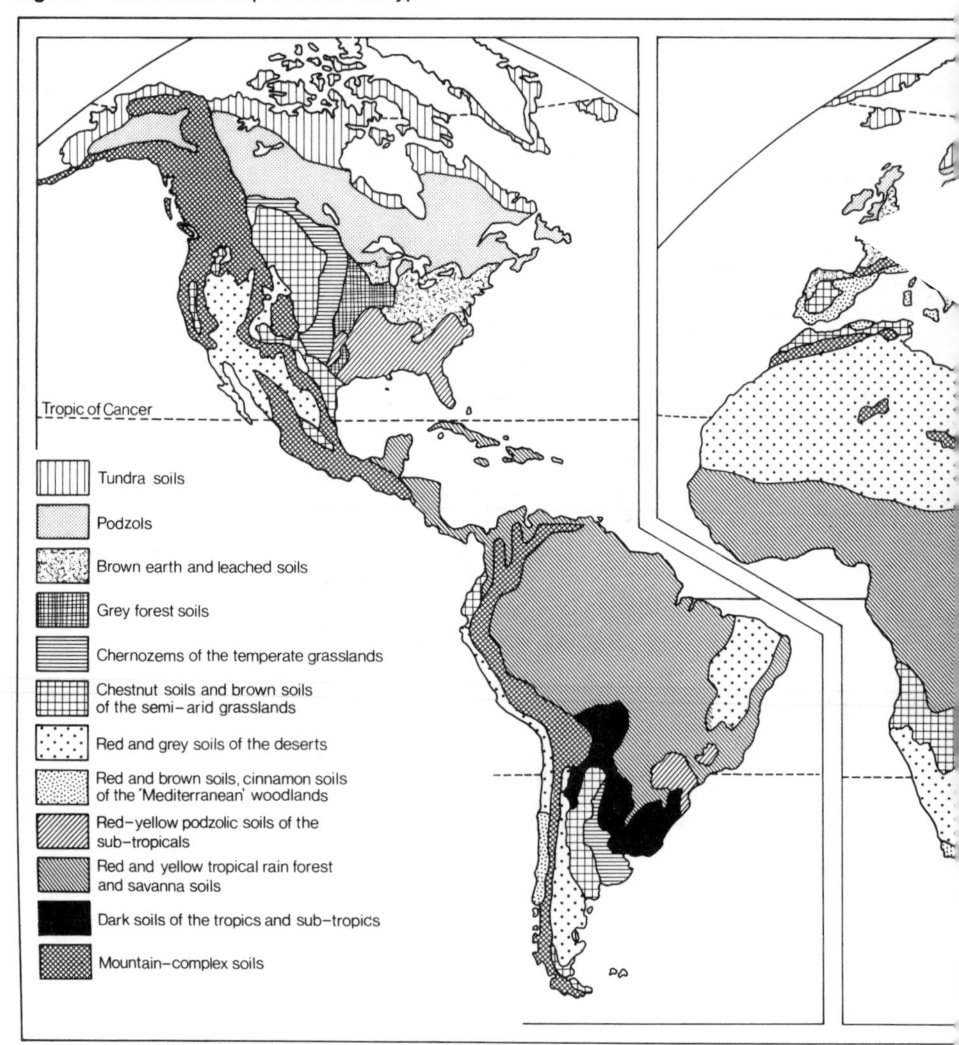

climatic extremes. A temperature range at Astrakhan of 24°C and only 140 mm of rainfall puts this site into the arid, Steppe climate which in the vastness of Asia becomes the cold desert region.

7. Moving yet further polewards from climatic types 5a, 5b and 6 can be found the boreal, cool temperate climate. This climatic type is poorly represented in the southern hemisphere because of the absence of land in latitudes 55°–60° S. In the northern hemisphere, comparable latitudes are the location for much of North America and the Soviet Union. Summers are characteristically cool with a July mean of about 15°C and a January mean of −8°C (both figures for Edmonton, Canada). Summers are moist with 300 mm of rainfall.
8. The final climatic zone is the arctic zone. There, seasonality is at a maximum with approximately 3 months being characterised by 20 hours/day of darkness, another 3 months with 20 hours/day of daylight and 6 transitional months.

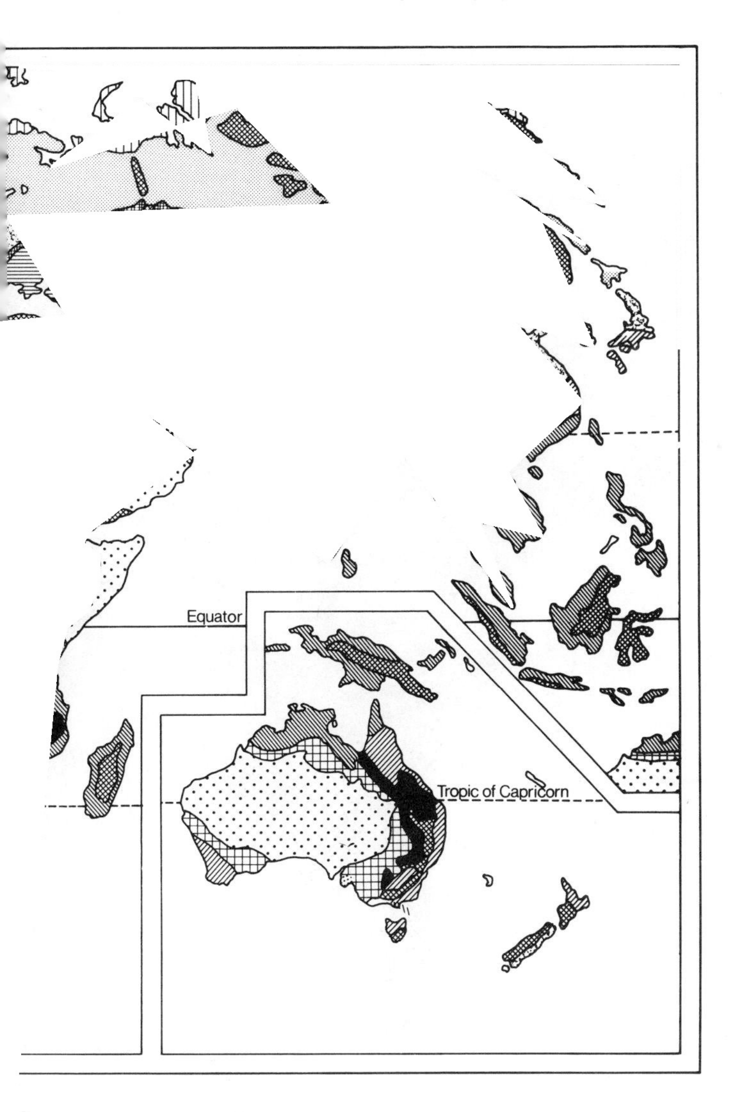

Such a standard description of climatic zones is obviously a gross over-simplification of reality. The effects of exposure and shelter brought about by mountain ranges, by local variations in the planetary wind system and by the modifying effects of warm and cold ocean currents cause the world climatic pattern to become a vast mozaic of sub-types. The climatic zones shown in Fig. 2.1 represent the broadest outline of conditions. The climatic classifications of Thornthwaite and Koppen recognise these broad types, but, as an illustration of the complexity of climate, they further define 32 and 31 main climatic sub-types respectively.

Even the most cursory examination of a world soil distribution map such as Fig. 2.2 will reveal a broadly similar pattern to that of world climatic distribution in Fig. 2.1. It was the Russian school of pedology which stressed the influence of climate upon soil evolution. In particular, Glinka (1931) was responsible for the view that under specific

Fig. 2.3 Generalised map of world vegetation types

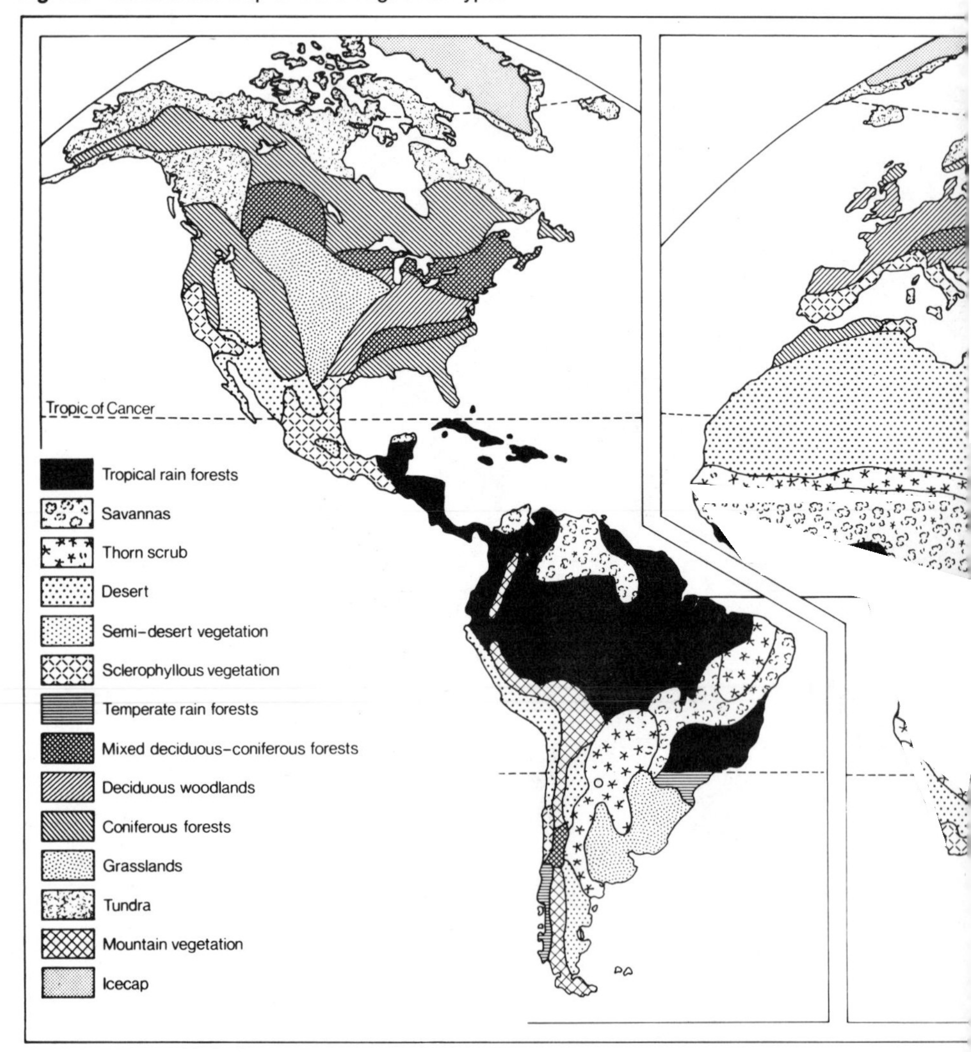

climatic conditions and irrespective of the original parent material then a specific soil type would come into being. Thus it was logical to argue that soil zones equivalent to climatic zones would exist. These were the so-called zonal soil types. Provided the concepts of the zonal school of soil zones are understood, as well as the limitations of the approach, then such a concept of soil forming processes does have a usefulness. It provides a global framework of soil types from the laterites of the equatorial climate, through the yellow-brown tropical soils, the Mediterranean *Terra Rossa*, the middle latitude brown-earth soils, the Chernozem and Prairie soils of the continental interior grasslands, the podzol of the boreal region and finally the tundra soils of the high latitude areas.

Opponents of the zonal approach argue that pedogenic processes operating within the soil profile are more relevant in soil evolution than the climatic parameters. The

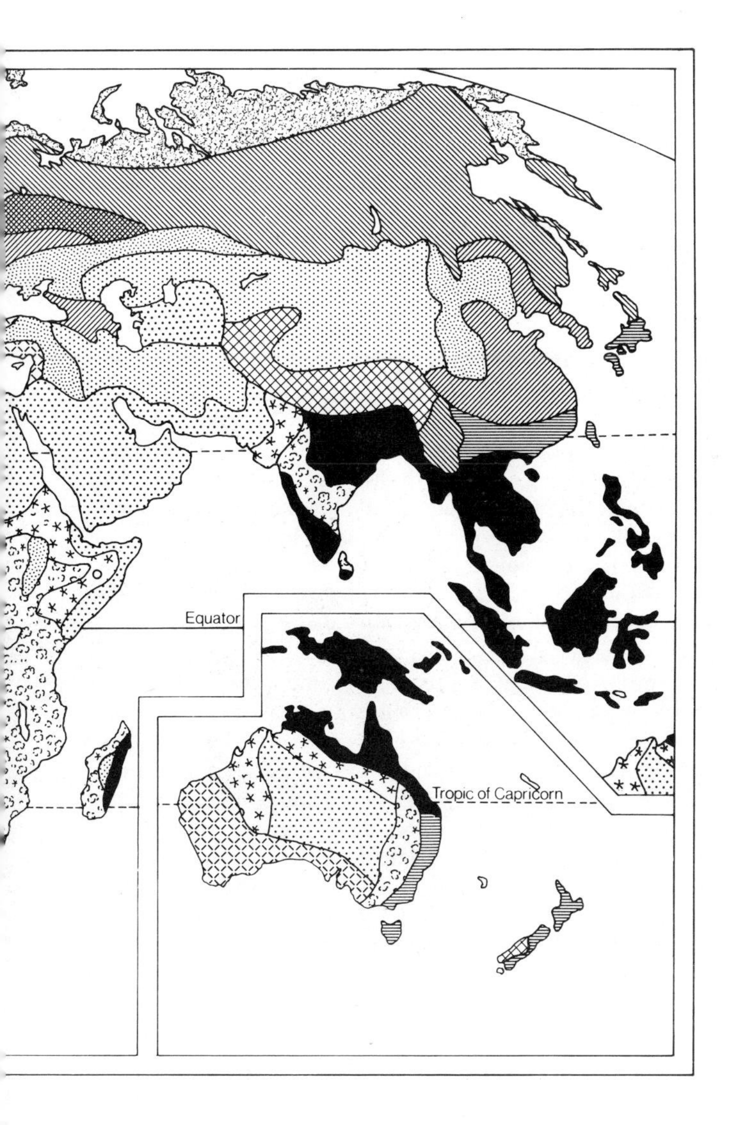

zonal soils approach demands that vast amounts of time are necessary for a soil to reach equilibrium with its environment. Climatologists can now show that climatic change has been operational at a much faster rate than was previously considered possible and it is unlikely that the soil would be able to evolve sufficiently rapidly and thereby match climatic change (Fitzpatrick, 1971). With these qualifications very much in mind it is necessary to stress that the combined influence of climate and soil does have very real control upon plant growth.

Following quite closely the general pattern of the world climate and soil regions are the major vegetation zones (Fig. 2.3). Let us now examine these zones, looking particularly at the natural productivity of the vegetation types. Again, a cautionary note is required. Most, if not all, of the vegetation on this planet has undergone some modification as a result of man's activities. Consequently, the vegetation types which remain are, at best, semi-natural units. Often the extent of the vegetation type is substantially reduced from what it was only a decade ago. There is evidence to suggest that the main impact of man upon vegetation types is that of simplifying the vegetation structure by removal of the species which have greatest economic value (Baker, 1970). Thus the remainder of this chapter discusses potential production which might be expected from the major vegetation types.

2.2 POTENTIAL PRODUCTION OF THE MAJOR WORLD VEGETATION TYPES

Rain forest

In this category can be found a vast array of forest types from the true non-seasonal, evergreen, tropical rain forest of which there are but small remnants in the Indo-Malayan peninsula to the degenerate, seasonal, i.e. deciduous, tropical forest of Colombia with its invasion of plantation-style agriculture.

Rain forest still covers a greater area of the earth than any other vegetation type, $17{\cdot}0 \times 10^6$ km^2, and it is inevitable that tremendous variation of shape, pattern, species composition and productivity will exist. Until quite recently, the bioclimatic conditions of rain forest were considered to be extremely constant and favourable to plant growth. These ideas have now been shown to be false (Janzen, 1975; Whitmore, 1975), particularly for the trees which form the over-storey or top layer of the forest. Extreme heating of the leaf surfaces may produce leaf temperatures of 10C° or even 15C° above ambient temperature and extreme water deficit occurs under these conditions. Tissue may dry out for periods of 5 or 6 hours before the temperature anomaly disappears. Under these conditions plant growth becomes erratic. Beneath the over-storey vegetation, however, can be found very different conditions for therein exists a dank environment with temperature and humidity conditions remaining almost constant. Temperature rarely falls below 25°C while relative humidity is commonly at or above 90 per cent. Lack of light is probably the limiting factor here, for the over-storey filters out at least 75 per cent of the potential illumination (Walter, 1973). Growth potential is shown wherever, and whenever, an old tree dies and falls to the ground. In temporary forest clearings the growth rate of bamboo has been shown to attain 300 mm/day. Trees may grow 35 m in 10 years. The problems of determining biomass and productivity of tropical rain forests are immense. Because the growing season is continuous, woody species do not produce the characteristic annual rings or annual growth scars which are common with temperate woody species. Age counts are thus impossible and the researcher has no convenient datum point from which to work. Again, any material which is harvested for analysis begins to decompose instantly because of the fecundity of the microbial activity. Weighing of material has other problems as the density of wood in rain forests can vary from S.G. 0·1 for cork wood, to S.G. 0·8 for iron wood, (Kira and Ogawa, 1971). The majority of rain forest biomass appears to be locked up in woody material, i.e. trunk, branches and twigs. Klinge (1975) recorded 97·4 per cent of the biomass of the Amazonian rainforest in the woody tissue category. Ovington (1965) provides biomass figures for Thai gallery forest in which branches, trunks and roots constitute 94·2 per cent of the total forest biomass. Non-woody tissue, i.e. leaves (5·2 per cent) and ground flora (0·05 per cent) were, on a percentage basis, unimportant. Total biomass figures produced by different workers for rainforest appear to be in general agreement. However, the same is not true for productivity data. Patterson (1956), Whittaker *et al.* (1973), Bazilivich *et al.* (1971) and Lieth (1973) all agree that the moist, tropical areas are the most productive regions, though comparison of their results is not easy. Indirect estimating, wide extrapolation and omission of below-ground material, all lead to difficulties in interpretation of results. For example, Lieth (1973) accords tropical rainforest with a total net primary productivity of $34{\cdot}0 \times 10^9$ tonnes p.a. while Bazilivich *et al.* (1971) assign the Tropical Belt, Humid Region with a net production of $77{\cdot}31 \times 10^9$ tonnes p.a. There is a substantial difference between these two values. Comparison is difficult, for not only do the two figures refer to different values – productivity and production – but the Russian source includes some vegetation types which are dubiously placed in the tropical rainforest category. For example, it includes tall-grass savanna, tropical mountain forest and flood plain formations all of which could be removed from the tropical rainforest category thereby reducing net production to $50{\cdot}21 \times 10^9$ tonnes p.a. The use of biomass values may help obviate the confusion. Biomass figures of between 450 and 750 tonnes/ha have been recorded (Lieth, 1973) though the upper figure, again of Soviet origin, appears too optimistic. Setting aside the controversy of which value is the most realistic, we must accept that even the lowest estimates of plant activity in tropical rainforest areas are still significantly greater than for any other vegetation unit on this planet. Table 2.2, located at the end of this chapter, gives the range of values for all the main vegetation types.

Whether judged on productivity, production or biomass values the rainforest areas represent the greatest producer of vegetable matter. The reasons for this are complex, but appear to be connected with the rapid cycling of nutrients between soil and vegetation, with the considerable species diversity and with the stability of the vegetation units.

Deciduous, low latitude forest and savanna

Moving away from the equator the equatorial climatic zone is replaced by an increasingly seasonal climate with distinct fluctuations in temperature. The rainy season coincides with the season of highest temperature. As a short, dry phase appears in the climatic regime then forest species begin to change. The top storey comprises deciduous and not evergreen species, though the understorey often remain evergreen. With a further decrease in rainfall and a lengthening of the dry season, then the proportion of deciduous trees increases. Only rarely can a single factor be identified as critical for tree growth; sometimes the amount of rainfall is the dominant factor, in others the length of the dry season, while in yet other instances the hot season or cool season controls the pattern of growth. More frequently, some or all of these factors combine and their total impact determines the growth conditions. Soil conditions too must be considered – the occurrence of laterite crusts, hard pans and stoniness are but a few of the possible controls on tree growth. Walter (1973) has suggested that only when the total annual rainfall drops below 500 mm does the soil type appear to become critical. A stony soil which receives less than 500 mm of precipitation is likely to be covered by a thorn bush type vegetation.

The effect of climatic seasonality upon vegetation productivity is profound, for no longer can plants grow continuously. Instead, growth processes must slow down, stop and then restart in harmony with the climatic regime. Inevitably, this interruption to the growth processes leads to a lowering of productivity, the extent of which depends upon the duration of the unfavourable season. Productivity will be reduced because of two factors: first, photosynthesis will be impossible during the dormant season; and second, a considerable proportion of new season metabolism will be absorbed in creating new leaf growth and not until this has been achieved can an increase of biomass really begin.

Net primary production (N.P.P.) for deciduous tropical forests varies from 400 to 2 500 g/m² p.a. depending upon the productivity controlling factors mentioned above. Lieth (1973) has suggested a mean N.P.P. value of 1 000 g/m² p.a. and a biomass value of between 420 and 460 tonnes/ha. More sources of data are available for savanna productivity. Once again values vary widely depending upon the duration of the dry season and the rainfall totals of the wet season. Lieth (1973) suggests an N.P.P. range of 200–2 000 g/m² p.a. while Bazilivich *et al.* (1971) provide production figures for a variety of savanna types. In order to compare the two sets of data the production figures have been converted to productivity values, thus:

Grass and shrub savanna	on	ferralitic red-brown soils	1 200 g/m² p.a.
Grass and shrub savanna	on	tropical black earths	1 100 g/m² p.a.
Grass and shrub savanna	on	tropical black earths	1 100 g/m² p.a.
Grass and shrub savanna	on	tropical solonetz soils	700 g/m² p.a.
Tall grass savanna	on	red ferralitic soils	1 600 g/m² p.a.
Tall grass savanna	on	black tropical soils	1 500 g/m² p.a.

Biomass data show similar accord between the two data sources with savanna showing a huge range from 20 tonnes/ha at the most arid margins to 600 tonnes/ha when the dormant season is minimal and tall grasses can flourish.

Subtropical, semi-desert and desert zones

In the zone of descending air approximately 23½° N. and S. of the equator can be encountered regions which experience potential evaporation rates which are much higher than annual precipitation levels. These are the subtropical desert and semi-desert regions. Temperatures can fall below 0°C during the cool season nights but rarely remain so for long. Lack of moisture is the overriding impediment for successful plant growth. Knowledge of desert vegetation has increased rapidly in the last decade and many of the previous concepts have been modified. For example, the paucity of vegetation in desert areas is known to be a natural response to available water supplies. The drier the area then the further apart the plants grow, thus providing a larger ground area from which the remaining plants can collect water. Competition between plants for water is thus minimised. When the transpiring surface area of desert plants is plotted against rainfall totals then a linear relationship is usually revealed. This pattern occurs over a wide rainfall range from 500–1 500 mm (Fig. 2.4). The supply of water to plants in deserts is not always as poor as is usually thought and when the xeromorphic

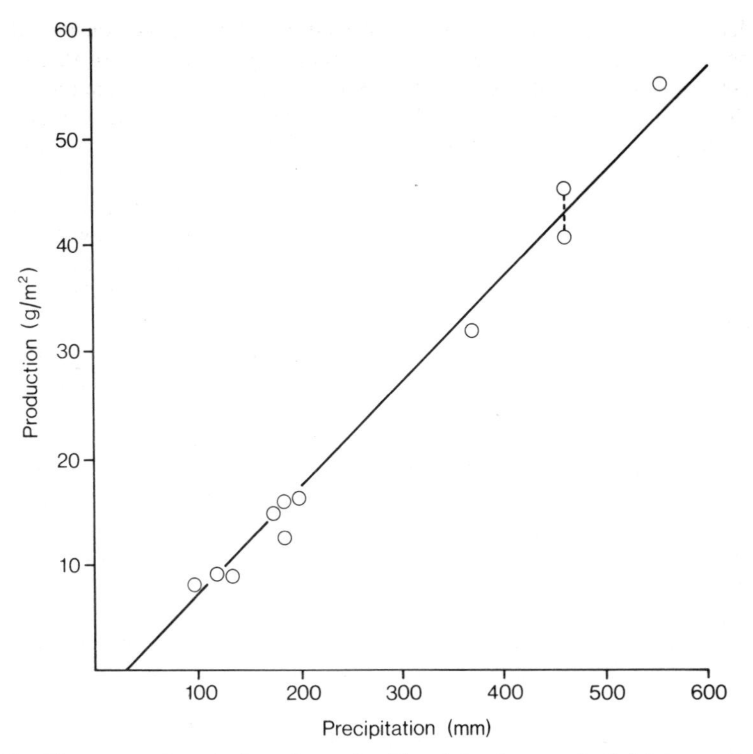

Fig. 2.4 Above ground production of grassland in relation to annual rainfall amount in South West Africa
(*Source:* from Walter, H., 1973, p. 87)

characteristics of the desert flora are taken into account it becomes apparent that conditions can exist whereby short, but extremely favourable growth periods exist.

The extreme scarcity of vegetation in deserts makes productivity values for the vegetation unit as a whole very low. True hot deserts are the least productive of all the world vegetation regions, attaining a maximum N.P.P. value of 10 g/m² p.a. and a biomass value of 3 tonnes/ha. These average, annual figures hide some very creditable growth rates when measured over days and not months or years. Desert tribesmen of the Negev desert were adept in practising 'run-off farming', a technique whereby small gullies would be dammed to retard run-off from flash floods and in which barley or even date trees could be grown. The natural vegetation of deserts appears to consist of a complex mix of perennial and ephemeral plants, the latter only germinating when an excess of water exists.

Within the spread of 23° of latitude can be found the extremes of global vegetation growth from the 450 tonnes/ha biomass of the Tropical Rain Forest to a meagre 3 tonnes (max.)/ha for the arid deserts. Almost without exception, man has failed to devise an agricultural system which can accommodate such profusion on the one hand and scarcity on the other.

Sclerophyllous winter rain zone (Mediterranean)

The aridity of the true desert and semi-desert zones ameliorates gradually as one moves polewards from the tropics. Summers are still dry, often very dry, but winters are characterised by cyclonic activity which can bring appreciable rain particularly in coastal or mountainous regions.

Distinct seasonality once again exists in the climatic regime and this is reflected in the vegetation type and in the growth patterns and hence productivity.

Typically, the natural vegetation was composed of evergreen hard and soft woods typified by *Quercus ilex* and members of the *Pinus* family. Very little of the original vegetation now exists due to the long human history of most of these areas. The native species show all the usual characteristics for survival in adverse conditions: sunken stomata, sclerotised leaf surfaces, reduction of transpiring surfaces, deep roots, and wide spacing between plants. Conditions during the winter time can be conducive to appreciable vegetation growth. The actual growth potential at a site depends very much upon detailed location, i.e. distance from sea, aspect, elevation. When a long dry summer is followed by a short wet season then productivity values as low as 250 g/m^2 p.a. may be recorded. In the western Mediterranean basin where winter rainfall is plentiful, then productivity can attain 1 600 g/m^2 p.a. Biomass values from a number of workers show variation, Bazilivich *et al.* (1971) quoting 170 tonnes/ha, Lieth (1973), 260 tonnes/ha, and Whittaker and Likens (1973) 350 tonnes/ha. Such a wide variation in biomass values is due to the critical nature of detailed location as already mentioned. Also, the impact of man is paramount in determining the detailed flora of Mediterranean lands. Where man still actively burns the vegetation and subsequently grazes his sheep and goats then the vegetation will be exceedingly impoverished comprising just single species such as *Quercus coccifera* or *Juniperus oxycedous.* It is inevitable that such a simple flora will show a low biomass value though the productivity rate may well be quite high due to the stimulus to growth created by firing.

Mid-latitude deciduous woodland zone

In the Northern hemisphere middle latitudes (45°–55° N.) can be found a temperate climate zone with a pronounced though not prolonged cold season. Such a zone is largely absent from the southern hemisphere except for small areas of New Zealand and the Southern Andes. The vegetation of the northern hemisphere mid-latitudes exhibits an obligatory shedding of leaves throughout the autumnal period and a production of new leaf material in the following spring. The deciduous nature of the vegetation is entirely due to the low temperatures in contrast to deciduous tropical vegetation which loses foliage in response to water deficiency. It is essential for the survival of the new growth in spring time that at least four months exist wherein photosynthesis and growth can occur thus allowing reserves to be accumulated in anticipation of the following growing season. Just as the Mediterranean zone has been subjected to extensive human interference the same also applies to the Mid Latitude zone in the northern hemisphere. Accordingly, reconstruction of former vegetation patterns and productivity values are difficult. Walter (1973) has suggested that the original forest cover has for so long been replaced by alternative vegetation types that forest may be genuinely regarded as being no longer the climax vegetation. Be that as it may, there remain enough forest remnants from which it is possible to reconstruct the productivity of the temperate deciduous forest biome. Productivity values drawn from throughout Western Europe suggest oak forests to be slightly more productive than beech while of critical significance is the state of the forest floor. Ovington (1965) has recorded between 3 000 and 4 000 tonnes/ha of litter production derived from old oak woodland and Stark (1972) has shown that an abundant litter and the presence of large numbers of litter-decomposing fungi have the ability to resist the ubiquitous leaching processes of mid- and high-latitude sites. As a general guide it would appear that the general luxuriance and species diversity of the mid-latitude deciduous forest may be considered a visual reflection of the inherent productivity of the ecosystem. Duvigneaud (1971) has shown that when the forest floor has a poor ground flora, or when the soil is waterlogged and/or has a low mineral status then the productivity value for the forest is low. An N.P.P. value of only 200 g/m² p.a. represents a production of very low proportions but can be recorded at upland sites with heavily leached soils. More typically, production values of 1 200–1 700 g/m² p.a. characterise the temperate deciduous forests, Whittaker *et al.* (1973) suggesting 1 200 g/m² p.a. to be a generally applicable representative value. Again, biomass values vary over a wide range in response to the fluctuating productivity values; biomass of between 101 and 384 tonnes/ha p.a. have been quoted by Duvigneaud (1971), while Whittaker has estimated 300 tonnes/ha p.a. as a mean value.

It is of relevance to discuss here the factors which can influence the productivity and biomass values of vegetation. Much emphasis has been given to climatic controls – the occurrence of distinct seasons based upon the lack of or availability of moisture and heat. Soil, too, will assist in the control of growth rate for a waterlogged site or an impoverished soil rarely proves capable of high productivity and high biomass values. Time is a factor which is often ignored in productivity assessments. Modern man is all too used to instant development. In nature the opposite is true. Imperceptible change with the passage of time is the characteristic of a balanced ecosystem. Great lengths of time are necessary for the development of plant associations, and of the plant layers within a forest. Natural processes rely very much on chance factors; for example, that seed is available from a parent tree or the chance that that seed might germinate and grow successfully without an animal eating it before it reaches maturity, sets seed and prepares the next generation of the species. Plants arrange themselves into groups, or associations, which have common habitat requirements. One of the commonest ways in which plants cooperate between themselves is in the formation of distinct strata or layers with each storey completing its critical part of the life cycle before the storey above begins its development. In this respect productivity of individual layers depends upon the so-called leaf area index (L.A.I.) or leaf surface area. The L.A.I. is defined as the ratio of the leaf surface area to 1 square metre of ground surface. It appears that the higher the L.A.I. value then the greater the potential for productivity.

There is little doubt that the multi-layered nature of virgin temperate deciduous forest facilitated species diversity and species productivity. As man simplified the forest ecosystem so he eliminated species and gradually caused whole layers to disappear from within the forest. For example, Pennington (1969) discusses the decline of the Elm tree, *Ulmus glabra*. Under such circumstances productivity and biomass values inevitably became reduced. The mid-latitude deciduous forests have been the home of man from about 5000 BC onwards. Forests were cleared to give way to agricultural land and cultivated crops now grow in place of natural forest flora. Chapter 3 will clearly show that even intensive agriculture cannot match the productiveness of natural deciduous forest. It can be assumed that the diversity of composition and structure of the natural vegetation unit permits the optimum vegetation productivity and biomass production that can possibly exist under the prevailing environmental and site conditions.

Coniferous forests (Taiga)

The vast zones of boreal forest extend across North America and northern Eurasia. Their extent today is not substantially different from that in the recent past for replanting of cut-over areas is now almost universal policy. Detailed floral composition has been considerably changed with emphasis being given to species such as spruces which can grow rapidly and are suitable for conversion to pulp.

Boreal forest extends over approximately $12{\cdot}0 \times 10^6$ km². It is inevitable that variation in climate, in soils and hence in vegetation composition should occur over such an extensive area. For example, along the southern boundary where coniferous and deciduous species grow as mixed forest either in alternating pure blocks or as a complete mixture, the cold winter season lasts for 6 months and the summer season has a maximum of 120 days when the average daily temperature attains 10°C. In contrast, the northern boundary merges with tundra vegetation and here a cold season of 8 months and a 'warm' season of but 30 days when average daily temperature attains 10°C is the norm. Temperature range can be as little as 30°C (−5° to +25°C) on the extreme western, oceanic coastline of Scotland and as much as 100°C (−70° to +30°C) beyond the Urals.

Throughout these vast areas may be found a vegetation type which exhibits extreme specialisation – the coniferous tree. Much argument exists as to whether the conifer is a specific adaptation to the inhospitable northern latitudes, or as Seddon (1971) has argued, that the features common to the majority of conifers are inherited characteristics of an ancient ancestor which permit these trees to exist within the cold, boreal climatic zone. Certainly, the *Coniferales* have their origins in the Triassic era 170–180 × 10^6 yrs ago, or earlier. It is true also that fossilised *Coniferales* are repeatedly found in areas now characterised by tropical or even equatorial climates. One possibility is that the *Coniferales* evolved in low latitudes and migrated polewards. Conversely, it is hard to imagine that chance alone should produce a vegetation type so perfectly suited to Boreal conditions. The list of characteristics which suit the conifer to its present-day habitat are impressive, viz., the shallow root pan, the mycorrhizal infections of the root hairs, the simple unbranched main stem and its associated downward sweeping branches and the magnificent photosynthesising organ – the rolled needle with numerous features to restrict respiration and water loss.

The characteristic of layering in forest vegetation types has been frequently referred to in earlier sections of this chapter. The northern coniferous forest is unique in that it does not exhibit this feature. There are a number of reasons which explain this fact. The major reason is the lack of solar energy; by the time the solar radiation has passed obliquely through the atmosphere and distributed itself over the surface of the earth there is relatively little energy available for photosynthesis. The evergreen habit of the conifers is just one way of optimising available energy input. Even in 'open' coniferous forest, that is where the trees are widely spaced, the trees cast a heavy shade upon the ground and consequently there is insufficient energy available for a luxuriant ground flora. Other factors which mitigate the formation of vegetation layers are the impoverished soils with low nutrient availability and the regular browsing of the vegetation by the nomadic reindeer herds. Walter (1973) gives figures to show that the soils beneath a spruce forest are 2C° cooler than soil temperature of open ground. This is due to the inability of the radiation to reach the forest floor. The temperature gradient is the opposite from that found under deciduous mid-latitude forest where the forest soil is invariably several Celsius degrees warmer than an adjacent non-forested soil. In the mid-latitude instance the forest acts as a retainer of heat while the high latitude forest repels what little radiation it has available. Under such austere conditions it is amazing that tree growth is at all possible. However, not only are there extensive forests but from the 1920s onwards man has actively begun to manage these forests and today they represent a major economic resource. There is little doubt that the reason why trees can grow in these high latitudes is due to the presence of a very efficient photosynthesising organ, the needle, a spiny 'leaf' in which the total surface area of the needle complement of a single tree far exceeds that of any broad-leaf angiosperm. This is reflected in the very high leaf area index values recorded for conifers. Walter (1973) has shown that the L.A.I. increases as harsher environmental conditions are encountered; thus at the southern margin of coniferous forest an L.A.I. of 9 is recorded in the pine forests while at the northern boundary only the most hardy spruces survive, now with an L.A.I. of 11. Lieth (1973) quotes an exceptional L.A.I. value of 16 for some coniferous species. In all probability the high L.A.I. values shown by conifers compensate to some extent for all the adverse variables which work against the growth of trees in boreal regions. Nevertheless, this compensation effect does not allow the conifer to be a massive producer of protoplasm. Lieth (1973) quotes net primary production to lie in the range 200–1 500 g/m² p.a. with a mean of 500 g/m² p.a. Whittaker and Likens (1973) have suggested a slightly higher mean of 800 g/m² p.a. while Bazilivich *et al.* (1971), give figures for eight different taiga, based mainly on soil differences, in which N.P.P. ranges from 400 g/m² p.a. for northern taiga on gley-permafrost soils to 1 000 g/m² p.a. for southern taiga on yellow-podzolic soils. Biomass values are, as ever, more complicated to calculate and compare than are production values for within the coniferous forest a large proportion of the biomass is locked up within the needle litter. Although conifers are labelled 'evergreen' they do in fact renew between 25–33 per cent of their needle complement each year. In old hemlock stands in western North America the quantity of litter from old stands can exceed 1 000 t/ha p.a. This does not all accumulate, however, and most of it, between 90 and 95 per cent, eventually decomposes to form peat. Walter (1973) has apportioned the biomass of an old pine forest accordingly: biomass of tree stratum 270 t/ha (max.), biomass of undergrowth (the meagre ground flora) 20 t/ha, total biomass 290 t/ha p.a. This is in agreement with a number of other results. Bazilivich *et al.* (1971), give a range from a low value of 80 t/ha p.a. for taiga on boggy sites to a maximum of 350 t/ha p.a. for southern taiga on yellow-podzol soils. North American data suggests a slightly higher biomass value, Lieth (1973) giving a biomass range of 200–520 t/ha p.a. and Whittaker and Likens (1973) 200 t/ha p.a.

In comparison with most other natural woodlands the northern coniferous forest is a low producer. A correct perspective of productivity and biomass values of the coniferous forest can only be obtained when the severity of the total boreal environment is recognised.

There remains only one environment in which the physical environment assumes an even more adverse factor to plant growth. This is the tundra zone. The climatic parameter which produces tundra conditions is a lack of heat. More accurately, tundra regions can be defined as having fewer than 188 days p.a. with a mean temperature of 0°C or above. Vast quantities of energy in summertime go, not to raising soil temperature, but in melting frozen snow. Wind exposure is also severe as a result of wind speeds of 15–30 m/sec in wintertime blowing away the protective cover of snow and ultimately ripping away the then unprotected tundra vegetation. The predominating vegetation is one of dwarf birth and willow growing amongst *Eriophorum* spp. (Cotton grass) and *Carex* spp. (Sedge). The steep, stony slopes that face south and west are often covered by a mozaic of small, but highly colourful annual plants which can successfully complete their life cycle within the short growing season but contribute little to the biomass values for the area. The size of the seeds of tundra species reflects the low productivity of the biome since 75 per cent of the seeds weigh less than 1 mg. Considerable work has been done on tundra productivity and biomass values (Wielgolaski, 1972; Webber, 1974). Because of the relatively sharp environmental gradients which exist from the tundra margins towards the poles it has been possible to correlate the decline in growth potential with productivity values (Table 2.1). Bazilivich *et al.* (1971), once again provide figures that are at variance with this sequence; the six tundra situations quoted by this source suggest that both productivity and biomass values are very much greater, thus:

polar deserts	100 g/m²	5 t/ha
tundra on tundra gley soil	250 g/m²	28 t/ha

Both Whittaker and Likens (1973) and Lieth (1973) give generalised tundra values of N.P.P. at 140 g/m² and biomass varying widely between 1 and 30 t/ha.

Table 2.1 Variation in tundra primary production values

	NET PRIMARY	
	Production (g/m²)	Biomass (t/ha)
Extreme arctic (growing season 60 days)	3	–
Northern tundra (cushion plants)	40	0·6
Central tundra (herb rich)	70	1·2
Alaskan tundra (growing season 70 days)	83	–
Southern tundra (dwarf willow and birch)	120	3·2
Tundra–Taiga transition zone	140	7·3

(*Source:* from Walter, 1973, pp. 213–14)

2.3 CONCLUSION

It is obvious from the productivity and biomass values given in this chapter that there is immense variation not only between the different vegetation units but also within vegetation types. This is not only confined to natural vegetation types but is also apparent in modern agricultural enterprises. For example, the effect of shelter has long been known to be beneficial to plant growth and Hogg (1959) presented figures to show that an artificially sheltered potato field gave a 21 per cent increase in crop yield when compared with an unsheltered field. It is almost certain that many micro-differences in soil and climatic factors cause fluctuations in productivity and biomass values. Within any growing season and at any site some of the factors will be growth promoters and others are growth retarders and over a long period they may balance the impact of each other. On a short time scale, however, some erratic growth patterns may be observed – for example, favourable spring-time weather can cause an early and rapid start to the growing season. For this reason care has to be taken over the timing and duration of productivity and biomass assessments. There is a general lack of long-term plant growth measurements (some forests and experimental farms excepted). Many of the production figures, particularly for the low-latitude natural vegetation types, are isolated values, the result of expeditions, or research projects which last for several months as opposed to years. Longer term assessments are almost totally lacking and thus we have little idea how productivity and biomass values, for the same area, vary between years. There is a dearth of reliable data for low latitude vegetation types while the middle and high latitude vegetation units have only recently been more thoroughly researched.

It is of paramount importance that we discover the potential protoplasm output of the different vegetation units, for if we accept that the natural vegetation of an area represents the ultimate vegetation development for that site then it follows that the natural vegetation unit will also possess the optimum productivity. In a world where organic resources are increasingly at a premium it is essential to know how much vegetable production man can expect to harvest from the remaining vegetation types on this planet. It is also necessary to know the optimum production levels so that man has a target to which he can direct his agricultural and/or afforestation policies.

Table 2.2 Summary chart of vegetation production for major vegetation units and agriculture (source of each value, where known, appears in parentheses ( ) following the production figure)

VEGETATION UNIT	NET PRIMARY PRODUCTION (g/m² p.a.)		BIOMASS (t/ha)		LEAF AREA INDEX m²/m²
	Range	Mean	Range	Mean	
Tropical rain forest	1 000–3 500 (1)		450– 750	450	8 (3)
	3 500–5 600 (2)	2 000	600– 800 (3)		6–16·6 (1)
	5 000 (3)				10 (8)
	5 000 (4)				
Seasonal rain forest	450–2 500 (1)	1 000	420– 460		
Savanna	200–2 000 (1)	700 (3)	2– 150 (3)		
	700–1 600 (5)		20– 600	366	
Arid desert zone	10– 250 (3)	70 (3)	1– 40 (3)	7	
'Mediterranean' zone	250–1 600		170 (5)		
			260 (1)		
			350 (3)		
Mid-latitude deciduous forest	200–1 200 (3)	600 (3)	101– 384 (6)		3·8 (3)
	1 200–1 700 (6)		60–2 000 (3)		5–8 (8)
	560–2 240 (2)	1 200–1 300	218 (7)	300	
	600–3 000 (3)				
Boreal	200–1 500 (1)	500 (1)	290 (8)	200	6·68 (9)
	400–1 000 (5)	800 (3)	80– 350 (5)		14·8 (3)
	350– 560 (2)		200– 520 (1)		9 (1)
	400–2 000 (3)		398 (3)		11 (1)
			270 (9)		16 (1)
Tundra	3– 140 (8)	140 (3)	5– 28 (5)	6	
	42 (2)		30 (1)		
	100– 250 (5)		1– 30 (3)		
	10– 400 (3)				
Agriculture	100–4 000 (3)	650 (3)	4– 120 (3)	10	

(1) Lieth: (2) Newbould: (3) Whittaker: (4) Kira and Ogawa: (5) Bazilivich: (6) Duvigneaud: (7) Dilys: (8) Walter: (9) Satoo.

CHAPTER 3
MAN'S UTILISATION OF VEGETATION PRODUCTION

3.1 MAN'S REQUIREMENTS FROM PLANTS

Man is bound by the same basic requirements as are all the other members of the animal kingdom. Above all else, he must consume food in order to survive. Having made that fundamental and obvious remark it is necessary to show that, other than the requirement to feed at regular intervals, man has developed very different feeding patterns from other organisms. Unlike most other animals, man is an omnivore and draws his food supply from a huge variety of types and sources. While most animals have alternative food supplies none are as diverse in their feeding habits as *Homo sapiens*. Man differs also from other animals in that he uses plants for a host of other functions besides consuming them; for example, in construction of dwellings, for medical supplies, for textile manufacture and for pulp and cellulose manufacture. Man differs mainly from the rest of the animal kingdom in his ability to choose the type of food he eats and from where he obtains the food supply relative to overall food chains. Modern man has a host of ancillary information to assist him plan an energy- and protein-balanced diet. Food additives can improve, retain or maintain flavour while canning, dehydrating, freezing or otherwise 'processing' of food can make food available in 'out of season' or 'out of location' situations.

3.2 ORGANISED AGRICULTURE

Contemporary agriculture and food processing industries have evolved considerably from early farming patterns. Just how the processes of plant and animal domestication began are lost in time though it is reasonable to suggest that man recognised, perhaps unconsciously so, that considerable improvements in his food supply could be achieved if he imposed some degree of management on the growth of plants and animals. It is considered likely that a prerequisite to the rapid increase in population which occurred in the late mesolithic and neolithic periods was the ability to develop a sedentary farming system with both plant and animal species being brought under domestication. Schwanitz (1967) has estimated that a hunting and gathering community requires about 20 km^2 of land per person in order to guarantee year-round survival. A similar area of fertile ground could, under a moderate agricultural system, support 6 000 people! The change from a collecting economy to an agricultural producing economy must have been a somewhat fitful, erratic process and one which it could be argued is not yet complete as shown, for example, by the 'collectors' in modern Western society who each autumn travel into the countryside to collect blackberries, blueberries, crowberries, hazelnuts, elderberries, etc. Watts (1971) has suggested that the beginnings of agricultural systems are most likely to have started on disturbed land surrounding the middens of the food gatherers and preferably set on a hilly site where environmental variables were likely to be constantly changing thereby increasing the possibility of natural hybridisation between diverse flora. Gradually, the food collector became aware that certain plants yielded more food than others, or that quality was superior or that roasting or baking improved the flavour of some species but not others. One is forced to speculate just how many of these early agriculturalists actually died from their involuntary experiments into primitive plant selection. For example, *Daucus carota* forms a fibrous, yellow root and appears a useful edible plant. Yet this species is deadly poisonous. Even more remarkable perhaps is the fact that *Daucus carota* is the ancestor to the carrot we now eat with little expectation of suffering death by poisoning! Hole and Flannery (1967) have presented sound evidence to suggest that the first steps in plant domestication occurred in the area now covered by western Iran at about 10 000 BP when emmer wheat (*Triticum diccocum*) and two-row hulled barley (*Hordeum distichum*) were grown. Another 2 500 years were required before the appearance of the modern wheats, e.g. *Triticum aestivum,* along with other species such as lentils, linseed and a proto-pea. Development of this system extended over a further 2 500 years, to 5 000 BP, before optimum sophistication was attained. Much controversy exists over where and how agricultural production first began. Why, for example, should the arid Iranian semi-desert be the location of early domestication of crops? Flannery (1969) has suggested that the arid areas of north-east Africa may be considered as 'over-spill' areas for the more humid tropical areas. In these humid areas the population carrying capacity for hunters and gatherers was high – perhaps too high, and an outward migration occurred, leading peoples into less suitable environments for hunting and gathering. As a consequence, innovation occurred, a direct result of which was the emergence of a more sedentary way of life based upon primitive plant agriculture.

Innovation has proved the hallmark of man's character. Man has been forced to innovate, to adapt, to experiment at every point of his existence as a distinct animal species. This is still very clearly shown today in modern agriculture where international trade policies, fixed prices and trade embargoes cause the modern farmer to be in the forefront of innovation processes. Tarrant (1975) has discussed the speed with which new farming technology and new or improved crop varieties can spread between farms in Britain. Using maize as an example, he has shown that innovation starts at a small number of sites (individual farmers) and then spreads outwards, relying much on transfer of information between farmers to convey the advantages and disadvantages of the crop. In complete contrast to the innovative farmers of the large, mechanised lowland farms are the traditional farms of remote 'back-woods' or upland regions typified by farms of Upland Britanny. A few small fields with only the essential elements of mechanisation, the use of well-tried methods, traditional plant and animal varieties all mitigate against the possibility of increased productivity. While it is easy to criticise the poor productivity record of this type of farm (and it is found in just about every nation in the world), it must be remembered that the farmer and his family are often victims of circumstance. As a result of the system of land inheritance whereby each son received an equal share of the land and assets (the process of *parcellement* in France), farm size was often reduced to little more than several fields. Physical factors are often the cause of poor productivity; upland, exposed sites with thin soils and short growing season cannot be rectified by government grants or fertilizer subsidies. Historical and social factors also loom large in any analysis of farming productivity patterns. In Britain, for example, agriculture as an employer of labour reached a maximum in the late nineteenth century. Approximately 90 per cent of the land surface was devoted, in one form or another, to agriculture (Best, 1968). Throughout the present century, agricultural land has been lost to a number of competing land users, notably to an increase in urban area (about 11·5 per cent of the land surface in 1977) and to afforestation projects (about 9·5 per cent in 1977) until land devoted to agriculture has fallen to 76·5 per cent of the land surface. Over the same time period the population of Britain has increased from $38{\cdot}2 \times 10^6$ in 1901 to 56×10^6 in 1971, a 47 per cent increase. This increase has created greater demand for food as has the increase in real income and associated standard of living. Britain has turned to overseas sources for many of her staple products, wheat being the notable example, and yet the home agricultural system has also been able to make massive improvements in productivity. Many other European countries have found themselves in a similar position to Britain; France, Belgium, Holland, Denmark all provide excellent examples of increasing agricultural productivity from a slowly declining agricultural area.

When one considers that man is almost totally dependent upon land-based agriculture it is perhaps surprising to note just how little land is devoted to agriculture in some continents. The F.A.O. Production Yearbook for 1977 shows that despite the highly agricultural nature of the countries mentioned in the previous paragraph only 25·0 per cent of Europe is devoted to arable agriculture. South America has the lowest value in the world with 6·8 per cent of the land devoted to arable, Africa 7·7 per cent, USSR 10·3 per cent, North and Central America 12·0 per cent and Asia 16·8 per cent. Only 11·2 per cent of the total world land surface is used for arable production, and from this fragment of world land area man obtains approximately 80 per cent of his food and organic materials used in processing industries. The remaining 20 per cent comes from aquatic sources, e.g. fish and seaweed. An ever increasing proportion of terrestrially produced organic matter is obtained from man-managed ecosystems, typified by highly organised farming and silvicultural systems. Not only are these systems man-made and man-maintained but the very pathways of energy and matter are regulated by man (see Fig. 3.1).

In the natural food chain the sole energy input is solar energy. This is cascaded through the system by the processes of plant photosynthesis and animal metabolism. At the junction of each trophic level, as energy is passed from one organism to another, there is an 'energy loss'. This takes the form of heat release, or energy used in respiration, locomotion and/or digestion. By the time the fourth trophic level is reached there is relatively little energy remaining and, normally, fewer individuals comprise each successively higher trophic level.

The man-modified agricultural chain is made far more complex by the additional inputs of energy to the system. This energy rarely takes the form of increased radiant energy (the exception being found in glasshouse cultivation) but instead is the addition of fertilizers, of manpower and of mechanised, i.e. tractor, power. This additional energy is required to improve the usable yield from the system. In the natural system much of the production of protoplasm is simply recycled by decomposer organisms and hence, in man's opinion, the yield of such a system is low. In the agricultural system the protoplasm production is channelled into ways which are of direct use to man. This channelling process is the fundamental aim of agricultural and forestry systems and when taken to its extreme, man controls the preparation of the soil by ploughing, draining and fertilizing, he controls the planting date, species and variety of crop, he dictates when any chemical spray should be added to control pests or disease or to accelerate ripening of the crop and he decides when and how the crop should be

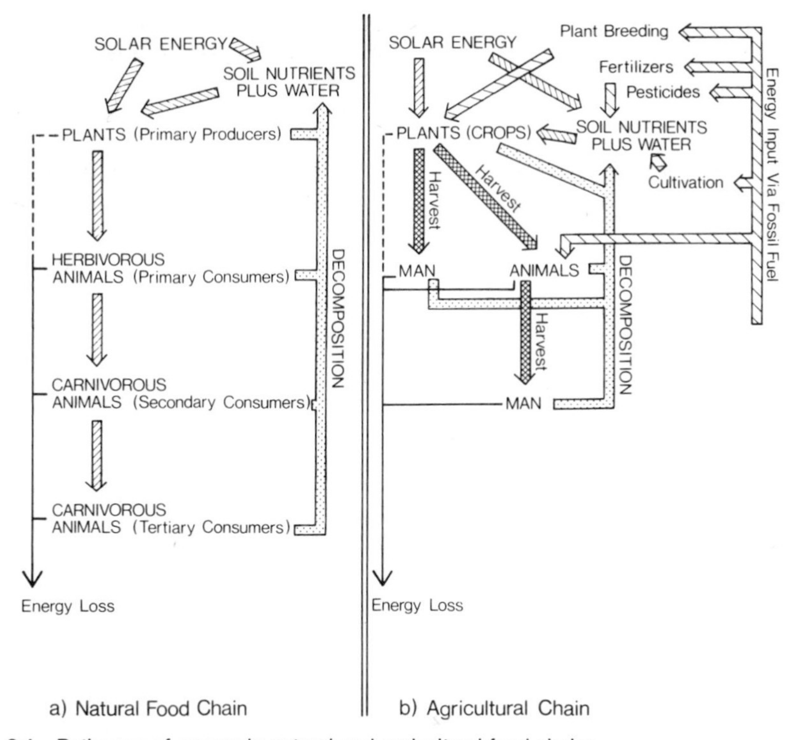

Fig. 3.1 Pathways of energy in natural and agricultural food chains

harvested. The implementation of this system demands that the farmer has the technology and finance whereby management processes may be executed. Implementation invariably means equipment and equipment involves the expenditure of energy to drive the equipment. There is considerable variation in the extent to which man has taken over the control of the natural food chain. In many African countries a common power input to the food chain has been from man's own muscles. This source of energy rarely exceeds the equivalent of 0·8 kcal/ha and the productivity of such a system is low. In a simple subsistence system in Uganda 3·2 ha of land was required simply to feed one family unit. By contrast, a highly mechanised farm 178 ha in size in England with a labour force of eight was able to produce sufficient food for 220 families (Duckham and Masfield, 1970). This enhanced productivity has to be paid for by an input of 2·4 kcal/ha for the mechanised farm compared with less than 0·8 kcal/ha for the subsistence farm.

Reference to Fig. 3.1b will show that man, as a feeding organism, can occupy varying positions in the food chain. In the simplest case, he can occupy the role of a herbivore and directly consume vegetation. Examples of such are commonplace: the rice-eating peoples of the Indo-Malayan region, or the potato-dependent population in Ireland, and to a lesser extent in Scotland, in the 1800s. Very high population densities can be supported if man is prepared to consume only vegetable matter. In these situations the energy losses are minimised because the food chain comprises but two trophic levels. Heavy dependence upon a single crop can lead to disastrous social consequences in years of famine, e.g. Ireland in 1846–47. Nutrition deficiencies can also arise from a single source diet. Duckham and Masfield (1970) give figures for the

Table 3.1 Energy efficiency of the potato crop in the UK

	MILLION CALORIES PER ACRE	PERCENTAGE
Total organic dry matter	24 750	100
Respiration losses	9 000	37
Unharvested vegetation (foliage)	3 750	15
Loss during storage	2 250	9
Household waste	2 250	9
Net balance remaining as human food source	7 500	30

(*Source:* from Duckham, A. N., and Masfield, G. B., 1970, p. 6)

energy efficiency of the potato crop in the UK (Table 3.1). Note that the role of the potato in the overall diet of man in the UK is now a very small one. These figures show that 70 per cent of the potential energy fixed by the potato plant and converted to chemical energy is never used by man. Indeed, 37 per cent is used up by the potato plant for essential growth processes. The 30 per cent of the photosynthesised material which reaches man may appear a disappointingly small proportion, yet in reality it represents one of the more efficient of man's food chains. A characteristic of Western civilisation is the appearance of large quantities of meat, poultry and dairy products in the dietary pattern. Most west European countries have between 42 and 44 per cent of the annual kilocalorie total originating from animal sources compared with 6 per cent for India, 23 per cent Greece. This, in terms of energy transformation through the food chain is an exceedingly expensive luxury which can usually be afforded only by societies with advanced agricultural systems and a wealthy, industrialised infrastructure. Turning again to Duckham and Masfield (1970) for data, it can be shown that an intensive grassland–beef cattle–man food chain transmits but 4 per cent of the potential photosynthate to man. The details of figures are given in Table 3.2.

Table 3.2 Energy efficiency of an intensive grassland–beef cattle–man system

	MILLION CALORIES PER ACRE	PERCENTAGE
Total organic dry matter	28 000	100
Lost by plant respiration	9 500	34
Lost in grass roots and unharvested stubble	3 000	11
Uneaten grass	4 000	14
Losses via animal excreta	4 000	14
Animal metabolism	5 000	17
Tissue conversion loss	1 000	4
Slaughter and household waste	500	2
Remainder for use as human food source	1 000	4

(*Source:* from Duckham, A. N., and Masfield, G. B., 1970, p. 7)

The intervention of a herbivore (the cow) in the food chain is to literally throw away 31 per cent of the available chemical energy which might have been available to man (14 per cent lost via animal excreta and 17 per cent lost via metabolic processes). Table 3.3

Table 3.3 A comparison of energy availability to mankind as produced from different agricultural processes

Arable system	Per cent
Potato crop and sugar beet	0·21–0·25
Cereals	0·16–0·20
Intensive grazing system	
Summer milk (experimental conds.)	0·15
Summer milk (normal production)	0·03–0·08
Beef rearing	0·005–0·025
Extensive grazing system	
Ranching	0·002–0·004
Fat Lamb production	0·001
Arable combined with indoor, intensive livestock system	
Pigs for bacon Barley – beef Broiler hens and eggs	0·015–0·030

(*Source:* from Duckham, A. N., and Masfield, G. B., 1970, p. 8)

presents, for comparative purposes, the photosynthetic energy availability figures for a range of farming systems. The lowly status of much of our agricultural production can be seen from this table. Much publicity has been given to the increasing gap between world population and available food supplies but there is little doubt that if the efficiency of agricultural systems could be improved, perhaps by as much as 10 per cent for the extensive grazing systems, then many of the food shortages would be alleviated – at least until world population once again caught up with agricultural output. At the present time the heavily populated developed countries (D.C.s) are dependent upon underdeveloped countries (U.D.C.s) for many essential vegetable products, e.g. sugar, rice, tea, coffee. There is little doubt that the D.C.s with their high level of economic development and high overall living standards also possess the highest level of food consumption and lowest levels of malnutrition. The F.A.O. have estimated an average 2 520 kcal/day energy intake for D.C.s compared with an average 1 990 kcal/day for U.D.C.s. The latter figure represents a 6 per cent deficiency in carbohydrates. Protein deficiencies are more serious, there being an average 7 per cent deficiency in U.D.C.s though shortages vary greatly from country to country, central Africa being poorest off with only 68 per cent of its protein requirement.

3.3 AGRICULTURAL V. SEMI-NATURAL VEGETATION PRODUCTION

It is of use to compare the net primary production (N.P.P.) of a man's agricultural systems with the selected natural or semi-natural vegetation units reviewed in Table 2.2. Net primary production values for agricultural crops have to be interpreted differently from natural vegetation data. The main considerations are the length of time the crop is in the ground and the extent to which the crop is 'assisted' by the addition of fertilizer. Intensity of farming methods are difficult to measure using simple indices but differing farming methods do produce very great differences in N.P.P. values. Instead of quoting single N.P.P. values it is better to quote world average values and intensive values, the latter serving to indicate the potential N.P.P. Table 3.4 gives production values for four crop types. The annual N.P.P. values when compared to equivalents for natural

Table 3.4 Net primary production values for selected agricultural crops

	N.P.P. VALUES (g/m²)		
AGRICULTURAL TYPE	per annum	per day over year	per day of growing season
Wheat, world average	344	0·94	2·3
intensive (Netherlands)	1 250	3·43	8·3
Rice, world average	497	1·36	2·7
intensive (Italy & Japan)	1 440	3·95	8·0
Sugar beet, world average	765	2·10	4·3
intensive (Netherlands)	1 470	4·03	8·2
Cane sugar, world average	1 725	4·73	4·7
intensive (Hawaii)	6 700	18·35	18·4

(*Source:* from Duckham, A. N., and Masfield, G. B., 1970, p. 10)

vegetation units appear disappointingly low. This is particularly so for world average figures which are dragged down by the disappointing results obtained from marginal subsistence type agriculture. Even the intensive figures only just reach mean N.P.P. values for natural vegetation types. Why, for example, should the intensive wheat farm in Holland show such poor annual N.P.P. values? The answer, quite simply, is that most of man's agricultural crops are annuals – they have a relatively short growing season during which they can accrue biomass. In contrast, natural vegetation offers perennial cover; the vegetation unit will comprise a variety of species with differing lengths of growing season. It is inevitable that such a mosaic of vegetation shows a better annual N.P.P. than a simple structured, one-species agricultural system. If we assess the latter in terms of N.P.P. per day of growing season (column 3 in Table 3.4), then far more favourable productivity values are obtained. Two conclusions emerge from this and will be examined in the following section.

3.4 POTENTIAL IMPROVEMENTS TO AGRICULTURAL PRODUCTION

Wherever possible, man should avoid monoculture and extensive fallow periods in the farming year. Fallow has traditionally been the recognised method of allowing the soil to regain fertility after an exhaustive period of cropping. No natural ecosystem requires a fallow period though it could be argued vegetation dormancy does represent a resting period. Fallow often, though not always, involves a lengthy period of time when the surface of the soil is bare or at best covered by a sparse cover crop of grasses and/or legumes. After an interval of 4 months or more the crop is ploughed back into the soil. Such a process cannot contribute directly to agricultural production. Indeed, a de-vegetated surface can undergo massive soil erosion as shown by Bormann and Likens (1970) in their experimental catchment area in New Hampshire, USA. Here, a natural forest ecosystem was completely clear felled and ground flora sprayed with growth-retarding chemicals in order to completely prevent regrowth. Although great care was taken not to break the ground surface, weathering and erosion increased at an alarming rate. Figure 3.2 shows the main trends which followed disruption of the ecosystem. The agricultural necessity for a fallow period has been strongly diminished now that greater understanding of crop rotation and plant food requirements have been obtained along with better control of fertilizer application and improved methods of soil cultivation.

Monoculture, however, remains a fundamental world-wide concept of modern agricultural practice. It represents the only practical way that man can produce crops on a large-scale basis. Multi-cropping (the growing of two or more crops mixed together or in distinct strata) or double or triple cropping (the production of more than one crop in a year from a site) are obvious methods of increasing N.P.P. values. Application of these ideas is far more difficult than at first appears. Double or triple cropping is possible in some locations. Rice has given the most dramatic increases in yield due to this method of growth; the trend began in the early 1960s in southern Japan, south China and parts of Formosa, Malaya and Indonesia. It depended upon the availability of new varieties of hybrid rice (a cross between two varieties of *Oryza glaberrima*, the '*indicas*' and '*japonicas*' varieties) developed at the International Rice Research Institute in the Philippines. While further advances in plant breeding are likely to make other crops suitable for double or triple cropping, it is unlikely that the geographical distribution of such a farming technique will be extended much beyond the limits of the tropics because of the lack of solar energy necessary for photosynthesis.

Multi-cropping holds out greater potential for improving N.P.P. values of agricultural systems. Research into grassland improvement has shown that grass–clover–vetch reseeding mixtures have considerably greater capacity to produce a highly productive sward ready for grazing than have straightforward grass swards. Little else has been achieved in multiple cropping although some barley–beef producers now undersow their grain crops with a grass–legume mixture so that when the grain is harvested the cattle can graze not only on the stubble but upon a nutritious green sward as well. This technique also delays the time when reploughing and reseeding is necessary and minimises the amount of time the field lies fallow. Scope also lies with the tree fruit growers – apples, plums, cherries, pears – for limited growth of early vegetables beneath the orchard trees. A common sight on the island of Mallorca is that of salad and

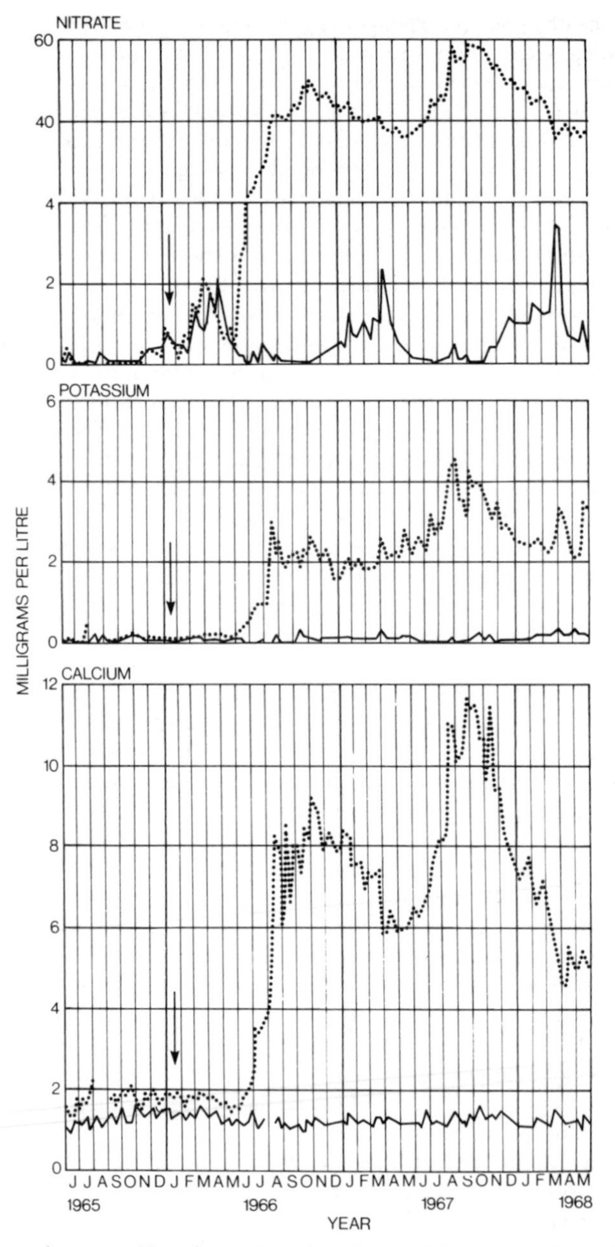

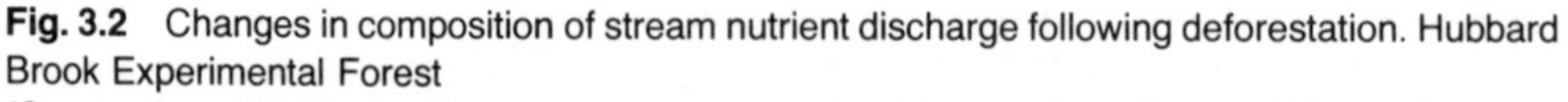

Fig. 3.2 Changes in composition of stream nutrient discharge following deforestation. Hubbard Brook Experimental Forest
(*Source:* from *The Nutrient Cycles of an Ecosystem* by Bormann, F. Herbert and Likens, Gene E., Scientific American Inc. © 1970, p. 100. All Rights Reserved)

vegetable crops grown with the aid of limited irrigation below the almond trees. In the counties of Kent and East and West Sussex in southern England, the massive inflation in production costs of orchard crops has led some growers to experiment with vegetable production beneath apple trees. Numerous problems exist, however, in these attempts at multi-cropping. Tree fruits need constant and careful spraying with chemicals in order to maintain high yield and blemish-free fruit. These additives are often totally inhibitive to the delicate, leafy vegetables growing at ground level. In addition, there are problems of harvesting date and harvesting technique. Each crop has either to mature at identical dates so allowing removal of both crops, or alternatively one crop has to be removed without causing damage to the remaining crop. Large-scale, field cultivation using multi-cropping techniques is unlikely to become possible until new varieties of plants are bred with features such as common harvesting dates or, for low growing species, the ability to grow under shade conditions. At present most of the common vegetable crops are strong light demanders and as such grow poorly when underplanted to taller species. Increased values of agricultural land along with a continued demand for more food from a smaller agricultural area are the factors likely to stimulate further research in this area.

A considerable improvement in agricultural production is likely to occur over the next 10 to 15 years although much of this improvement is likely to be due to improved farm management techniques and better storage of harvested goods rather than increased N.P.P. from crops or more efficient transfer of energy from one trophic level to the next. Indeed, Edwards and Wibberley (1971) have suggested a rise in Britain's agricultural output far into the future despite a gradual decline in land devoted to agriculture. It is hoped that such a prediction is correct and indeed that the trend will be world-wide so that the increase in world population will be matched by an adequate rise in food supply. Edwards and Wibberley (1971) consider that the demand for food is a function of three variables, viz. population size, the composition of population in terms of consumer units and the real wealth of the population. Unfortunately, there is no simple way in which these three factors can be combined to give a good production demand index figure. For example, it is the increase in population growth which is the dominant variable in determining food demand; it explains about 75 per cent of the increase in demand for food. It has become exceedingly difficult to forecast the way in which population is likely to increase even over a relatively short time span. Population increase throughout the 1960s had been occurring in the United Kingdom at about 0·7 per cent p.a. compound. Then, somewhat unexpectedly, the rate of increase declined until in 1976 rate of increase was 0·1 per cent. If this trend is maintained for several years then the medium term estimates of population size and hence food demand for Britain will be very much less than was expected.

Very few countries can now be considered to be self sufficient in terms of food production. Multi-national food companies have established subsidiary concerns in most continents so creating an 'international' menu instead of the considerable regional variations found until the 1950s. Some countries could revert to total self-sufficiency with little effort, for example, France and the USA. The more densely populated countries such as Britain, Denmark and Holland would have much greater difficulty. Britain, for example, was only 60 per cent self-sufficient in the temperate food products which she consumed in 1965. Increased productivity has lifted this to about 65 per cent (1975) while a conceivable level of self sufficiency for Britain might be 80 per cent by the year 2000. Perhaps the greatest stimulus to improved home production is the uncertainty of future policies regarding food production from the developing nations of the world. Increased home demand from producer nations often results in a smaller export potential. Occasional crop failures, trade embargoes, political revolutions, stock-piling, all can cause chaos for the food-importing countries, as instanced by the sugar shortage experienced by Britain in 1974. Often there are no alternative food sources, though to offset the cane sugar deficiency British farmers were encouraged to increase their acreage of sugar beet. Food shortages have stimulated the search for alternative, novel food supplies. The chemical compound, urea, is increasingly used as an animal foodstuff while recent advances have been made in yeast cell culture and other single-cell proteins have been grown on petroleum. Pyke (1971) has reviewed these new sources of food supply. At present high production costs and consumer resistance appear the main drawbacks to these synthesised food sources.

3.5 PRODUCTION FROM SILVICULTURE

The one other terrestrial source of plant production, beside agriculture, which man has sought to bring under his control has been that of the forest ecosystem. For most of man's existence on this planet he has been an exploiter of forest resources; by a combination of felling, burning, grazing of animals and collecting of economically valuable forest plants the extent of the forest cover has been very substantially reduced. Just how extensive the forest cover once was is exceedingly difficult to judge because of man's thorough efforts to clear it away. Of the total planetary land area ($13\,075 \times 10^6$ ha) the Food and Agricultural Organisation (1977) has estimated that $4\,156 \times 10^6$ ha (32 per cent) were covered by forest of one type or another in 1975. The proportion of forested land on each continent and in each country varies greatly (Fig. 3.3). In Britain the extent of afforested land has varied in post-glacial time, reaching an all-time high in the Atlantic period (5500–3000 BC) when an estimated 90 per cent of the mainland of Britain was covered by trees. The clearances which occurred first at the hands of farmers and later by industrialists were so thorough that by 1918 Britain possessed but 2 per cent of her land surface in forest cover (Edlin, 1956). France for example retains 20 per cent of her land surface under forest (Zuckerman, 1957) while other countries, notably New Zealand, have been deforested in a fraction of the time taken to clear the British Isles, 120 years compared with about 5 000 years! (New Zealand Forest Service, 1970).

Forests have, for long periods of man's history, been regarded as highly valuable resources. The Domesday Book of England makes reference to the wealth of a man being judged by the amount of forest at his disposal for the use of pannage by his pigs (Darby, 1977). This value of forests as a source of grazing area and/or materials implies an economic value and not a source of primary production and/or biomass. Ford (1971) has stressed that while the efficiency of modern forests in terms of growth rate is of paramount importance, it is valueless to compare primary production, productivity or biomass values between forests and agricultural crops. There are several reasons why such comparison is pointless. First the forest does not contribute directly to man's food chain, although experiments with extraction of protein from tree leaves shows this to be a feasible protein source should there ever be a major shortage of protein from alternative, cheaper sources. Second, the vertical and horizontal structure of the forest is totally different from the structure of an agricultural crop. The forest has considerable vertical extent (often 30 m or more) and usually occupies at least several hectares, and often many thousands of hectares, in extent. Trees are also perennial species and as such grow *in situ* for very many years. As they increase their stature so their impact upon their site changes, the interception of solar radiation and precipitation for example being directly related to the age of the stand. Ovington (1965) has shown that a beech forest in Switzerland intercepted only 2 per cent of the precipitation when 20 years of age, 27 per cent at 50 years of age and 17 per cent at 90 years of age. Agricultural crops rarely spend longer than 6 months in the ground though plantation agriculture is an exception to this.

Because of the immense structural, temporal and spatial differences between

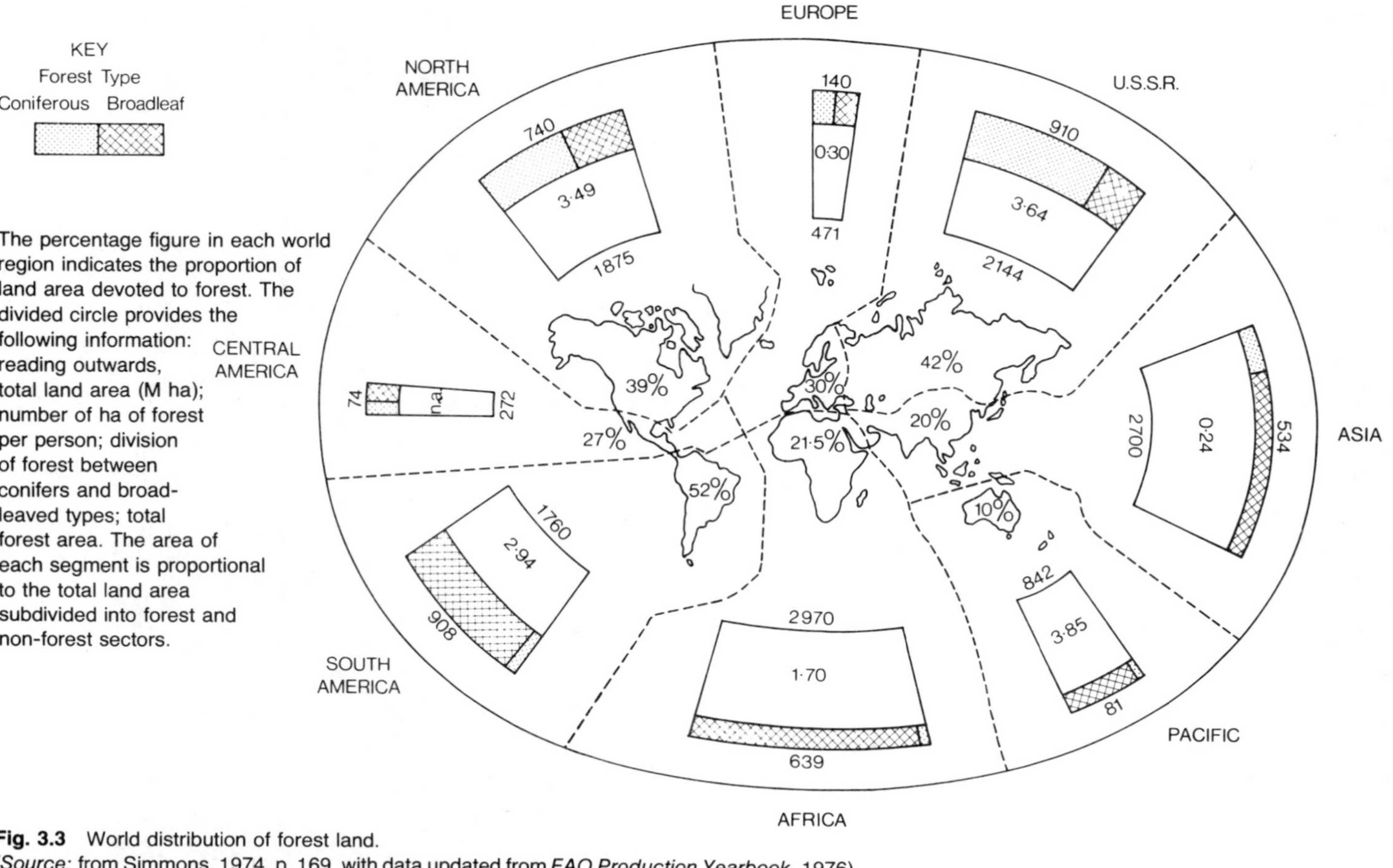

Fig. 3.3 World distribution of forest land.
(*Source:* from Simmons, 1974, p. 169, with data updated from *FAO Production Yearbook*, 1976)

forests and all other vegetation systems any comparison of productivity and/or biomass values is fraught with difficulty. Economic analysis provides an alternative method for comparison of forest and agricultural production and professional foresters much prefer this approach to a comparison of productivity values, partly because of the high value of timber! However, forestry is an extensive user of land resources and increasingly finds itself in competition with a variety of alternative land users. Forestry is thus gradually changing its character from that of an extensive, low-production land user to a less extensive, high volume producer of timber. Production level has become a key parameter in measuring the cost effectiveness of the forest but again production values are used not as the biogeographer or ecologist would use the date to indicate efficiency of the ecosystem. Instead, the forester uses production data to justify expenditure on items such as forest drains and roads, or upon the effectiveness of fertilizer application or ultimately upon deciding when to fell the forest and realise the capital investment. Thus commercial forest production data is used to justify further capital investment in the forestry system and only indirectly is it used to monitor the rate of tree growth.

The measurement of forest productivity is totally different from that of all other vegetation types for each year trees add to their volume by means of a further layer of woody material. Unfortunately, the annual increment of wood does not represent the true yield value for there will also be additional leaves, twigs and rootlets to add to the total productivity value. These latter items are of vital relevance to the ecologist's determination of forest productivity but to the professional forester they represent waste material which will be allowed to remain on the forest floor subsequent to the harvesting of the woody trunk. A more worrying possibility of modern afforestation schemes is the removal of the trunks from the growing site and by so doing carrying with them a proportion of the total incident energy for the production of the woody material. Research in German forests by Polster (1961) allowed him to allocate the following percentages of assimilated matter to the various components of a woodland:

45% lost to respiration processes
32% contained in trunk and main branches
16% in leaves/needles and litter
3% in roots
1% utilised in seed production

The remaining 3 per cent is destroyed during the processes of felling and removal of timber from the forest although the same author has suggested this last figure can rise to a massive 33 per cent when poor management allows wasteful methods to predominate. It is interesting to compare the figure of 32 per cent energy in harvestable timber with the figures given earlier in this chapter of 30 per cent for potato and 4 per cent cattle systems. Modern commercial forestry appears to compare well with agricultural practice in terms of efficient channelling of energy although the impact of deforestation on the mineral status of the soil may be less favourable. Reference to Fig. 3.2 will reinforce the comments made in section 3.4 on disruption of ecosystems. Timber removal inevitably causes severe ecosystem disruption and it is possible that the combined effects of timber removal and increased run-off and soil erosion could cause minerals which are already in short supply to become positive limiting factors in any future attempt to regenerate the forest site. Foresters will be particularly concerned in the future if tree growth rate and production amount begins to decline, thus suggesting that site degradation has occurred due to repeated removal of timber. Accurate recording of forest productivity and production is therefore a necessary requirement in present-day forest management.

CHAPTER 4
METHODS FOR ASSESSING FOREST PRODUCTIVITY

4.1 THE PURPOSE OF PRODUCTIVITY ASSESSMENTS

Little has so far been said about the actual measurement of vegetation productivity, production or biomass values. The next three chapters are devoted to methodology. Forest ecosystems have been chosen to illustrate methodology mainly because they provide a complete range of structural types, from the complex, multi-specied natural forest through to the mono-culture, 'anthropogenic forests' which now extend over so much of the northern latitudes and which are also appearing within the tropics. As a contrast to the measurement of forest production, Chapter 6 will show some of the special considerations necessary when assessing grassland and agricultural production.

The technique eventually chosen to measure productivity will depend upon two considerations. First, the type of vegetation unit – forest or grassland, natural or managed – will determine to a large extent the type of methodology which can be used. Second, the purpose of the survey can determine methodology. This is particularly so when the vegetation unit is a managed area when man, in his capacity as a farmer or forester, wishes to know the yield and the profitability of an area of vegetation.

No matter which technique is chosen to measure productivity, it is strongly advisable that a pilot survey is made in order to test the method for shortcomings. When working with a multi-specied forest the initial complexity can prove daunting; knowing where to start is a major difficulty though this may be overcome by examining just one species type at a time and gradually accumulating a comprehensive cover for all species types. If the results of a survey are to be compared with previously published results it is essential that methodology is similar and that comparison occurs between data with a common derivation. Sometimes a research project may wish to investigate the productivity of similar vegetation types but in differing geographical areas. On these occasions it is necessary to identify and control as many of the independent variables as possible between the sites. Thus variables such as elevation, exposure, soil type, aspect and management policy or historical evolution must all be known, and furthermore, must be similar between the sites under consideration. Such absolute control on the environmental variables is exceedingly difficult to achieve and often a compromise situation has to be accepted wherein the variables are reduced in number thus simplifying the choice of sites. Alternatively, modern quantitative methodology may be used though techniques which can accommodate multicolinear (interdependent) data are still poorly developed.

Despite the widely differing objectives of productivity assessments conducted by biogeographers, ecologists and professional foresters or agriculturalists, there is a very definite common base for all productivity surveys, that of the experiment design.

4.2 EXPERIMENT DESIGN FOR PRODUCTIVITY SURVEYS

A fundamental, but often overlooked, initial stage in productivity work is selecting an area and methodology which is within the financial and manpower capabilities of the worker. Measurement of vegetation productivity is extremely consumptive of labour and time and few relevant results can be achieved from forest productivity research in less than three years. A natural forest will be subject to fluctuating production dependent upon the interaction of the abiotic, i.e. physical, components and the biotic, i.e. organic, components of an ecosystem. Accordingly, average values made over at least three years are required. Exotic forests by contrast (that is, forests composed of non-native species such as the conifers in British forestry practice) tend to occur in even aged stands and tend to undergo distinct phases of production. At first they grow rapidly, reaching a maximum growth level after some 50–70 years before a decline in growth rate is shown. Thus, depending upon the age of the forest very different sets of production values could be obtained and the only accurate value for annual production would be the mean production value based upon the total life of that forest. This might involve a time span of at least 70 years, possibly more than 100 years. An alternative to this approach is to measure annual production at a number of sites with similar species but of varying age and to use this data as an approximate average value of production. Such a method works only if care has been taken to minimise the effects of varying independent variables as explained at the beginning of this chapter.

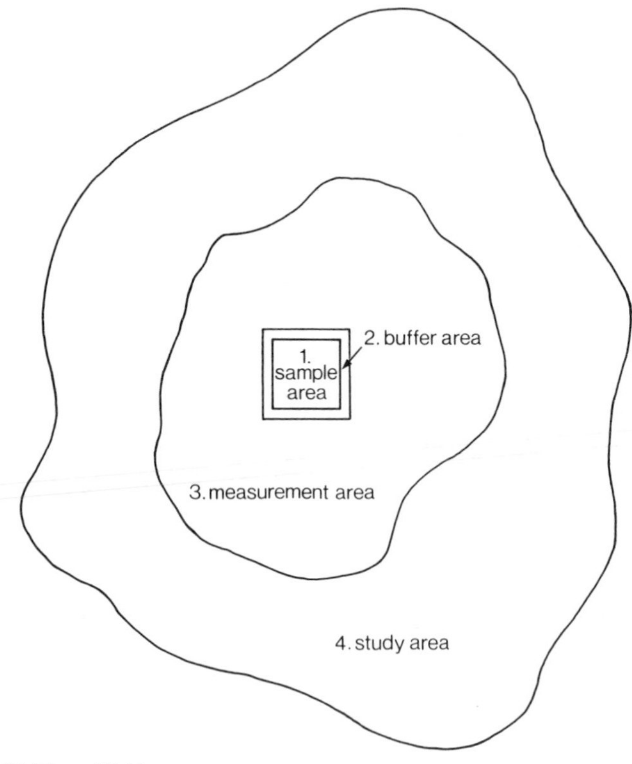

Fig. 4.1 Division of field area
(*Source: Methods for Estimating the Primary Production of Forests IB2*, see Newbould, 1970, p. 4)

Selection of a site suitable for the survey is of critical importance and above all else it must be representative of the whole forest area. Newbould (1970) has suggested a fourfold division of site as shown in Fig. 4.1. The innermost sample area provides a site for experimentation which will not change the natural growth pattern. Thus repeated growth measurements, but ideally not cutting nor the removal of any material, may be made in the sample area. Provided the sample area is judged to be representative of the whole site, then it can be as small as 0·1 ha. This 'inner sanctum' should be surrounded by a protective zone or buffer area at least two tree heights in width. Next comes the measurement zone in which the main destructive and non-repeatable measurements are to be made. Thus individual trees can be felled and removed for detailed biomass assessment, soil pits dug and tree root examination made. This area may have to be as large as 10 ha in order that all soil types and all tree species are included. Surrounding the measurement area is the study area, the function of which is to provide another buffer zone for the whole experiment.

Absolute standardisation of site layout as indicated in Fig. 4.1 is not always possible, nor indeed desirable. Replication of sites often makes the pattern shown in Fig. 4.1 impossible as overlap of zones would complicate the results. However, no matter what type of forest is being studied it is vital to embark upon careful experimental design before attempting to measure growth rate. Because of the great differences which exist between native and planted forests in terms of their structure, the species composition and their commercial usefulness then methodology relevant to the two main forest types will be dealt with separately.

4.3 THE DETERMINATION OF PRODUCTION FOR NATURAL AND SEMI-NATURAL FORESTS

The area to be assessed for production must be thoroughly reconnoitred and the vegetation assemblage divided into convenient units each of which can be investigated separately. Components such as dominant trees, under-storey trees, shrubs, herbs and ground layer vegetation are the most likely units to prove useful. It is improbable that each of these units will be equally assessed for production for the importance of each will vary between ecosystems while difficulty in assessing, for example, root systems of large trees may preclude an accurate record of this unit. It will be also necessary to decide at the design stage which of the various growth rate indices will be adopted. The differences between 'biomass', 'productivity' and 'production' have been explained in section 1.5. For natural or semi-natural forests the majority of published work has been made upon the net primary production values, that is the assimilation of organic matter by a plant community during a specified interval of time (usually a growing season, or one calendar year), minus the matter used up by respiration processes.

The most fundamental stages in recording the net primary production (N.P.P.) value are given by Newbould (1970) as the measurement of the 'net assimilation' (net dry matter production of green parts) over the year(s) or vegetation period(s). This can be determined by the sum total of the following features determined periodically through the year:

(*a*) biomass change of photosynthetic plants;
(*b*) plant losses by death and the shedding of parts above and below ground;
(*c*) man's harvest (in some cases);
(*d*) consumption of photosynthesizing plants by animals (botanical and zoological methods will be used to estimate amount lost).

An approach such as that outlined above can provide an accurate index of N.P.P. value though there will inevitably be some problems, the most obvious being the need to measure biomass of each organism at two distinct stages, usually the beginning and end of the project. Kira and Shidei (1967) have provided an alternative method in which N.P.P. is described as a mathematical function. In its simplest form N.P.P. can be estimated from the following equation: $Pn = \Delta B + L + G$

where: Pn = net production by the community during time t_1 to t_2

$\Delta B = B_2 - B_1$ = biomass change during period $t_1 - t_2$

B_1 = biomass of plant community at time t_1 (beginning)

B_2 = biomass of plant community at time t_2 (end)

L = plant losses by death and shedding during $t_1 - t_2$

G = plant losses by consumer organisms as herbivorous animals, parasites, etc. during $t_1 - t_2$

This equation still requires a double calculation of biomass, viz. at times t_1 and t_2. A major improvement can sometimes be made by harvesting vegetation at time t_2 and by measuring the current year's growth, an estimate of one-year-old organic material can be made. When Pn is calculated by this method it is often difficult to allocate material to specific periods of production while the assessment of dead plant material (L) and consumption by consumer organisms (G) for current year and previous years is sometimes also difficult. The biomass obtained via this method can be described by $(Pn - L_N - G_N) = B_{2N}$ = the apparent growth increment, while net production is $Pn = B_{2N} + L_N + G_N$ where L_N is plant loss from Pn during time $t_1 - t_2$ due to death and shedding and G_N is plant loss from Pn due to consumer organisms.

A sampling programme for the vegetation units has to be devised whereby non-destructive measurements such as height of vegetation, girth (or more accurately, diameter at breast height, D.B.H.) of trees, length of apical shoot, quality and type of litter fall all can be measured within the innermost sample area (Fig. 4.1). Destructive measurements, such as felling of trees, clip quadrat collections, current leaf quantity as measured by fresh weight, volume or ash drying must also be made but are confined to area 3 on Fig. 4.1. The aim of both the destructive and non-destructive sampling is to obtain a correlation between the sampled area and the remainder of the study area. If the forest is under any form of management then much statistical data may have been collected by the forester though, as has already been shown, it may be in a form unsuitable for use in ecological productivity surveys.

Accurate estimates of vegetation growth rate require precise measurement, for errors on a small sample become cumulative and when applied to a large area the final productivity/production figure can easily become invalid. Because of the immensity of complete vegetation surveys some workers have concentrated on specific parts of ecosystems. Madgwick (1970) has examined the biomass and productivity of forest canopies as he argues that the foliage is the producer of photosynthesised products and in turn is responsible for much of the dry matter production within the system. Two main methods for estimating biomass of foliage and branches are discussed by Madgwick (1970). It can be assumed that the tree has 'average dimensions', i.e. that the forest is composed of a variety of shapes, sizes and weights of trees which when summed and divided by the sample size will give average values. It is unlikely, however, that a tree which approximates, for example, the average basal area will also approximate any other average value. Thus the average dimension concept is not an infallible method for calculating canopy biomass. An alternative statistical method exists whereby a number of trees can be accurately measured to provide a regression equation for logarithmic canopy weight (log W) on logarithmic stem diameter breast height (log D) for estimating the weight of branches and leaves in the remainder of the stand. The regression equation may be written as follows:

$$\log W = a + b \log D$$

Values a and b are constants though Satoo (1965) has shown that these values will vary for each tree species depending upon the level of competition within the forest. Bunce (1968), working in a mixed forest in the English Lake District, has shown that, provided a wide range of different tree girths are available, then the same equation can be used for a multi-species forest.

The statistical modelling of plant growth is still only poorly developed yet its advantages are clearly apparent. Model plants can be 'rearranged' so that conditions conducive to optimum growth can be obtained. Also, it allows the impact of environmental variables to be assessed and can suggest changes in management patterns so that productivity may be enhanced.

A more traditional approach is required for the assessment of biomass in the main trunk of the tree. As the trunk (alternatively called main stem or bole) is a major component of the forest, it requires careful measurement. The measurement of the trunk can be made in one of the following ways (arranged in order of declining accuracy).

1. If the stand is part of a well managed forest then it will inevitably have been measured for stem volume on at least one occasion in its life history. Under such conditions the volume increment for the stems can be obtained by measuring the D.B.H. and/or top height and by reference to published volume tables for the species concerned, the volume can be readily calculated. Timber volume can be converted to dry weight by multiplying volume by a conversion factor based upon the appropriate specific gravity (Table 4.1).

Table 4.1 Average conversion values for freshly felled forest timber (m^3) to oven-dry wood weight (kg)

SOFTWOOD CONIFERS		HARDWOOD DECIDUOUS	
Weymouth pine	320	Alder	430
Silver fir	370	Elm	460
Norway spruce	390	Birch	510
Douglas fir	420	Maple	540
Scots pine	420	Beech	560
European larch	470	Oak	570
		Ash	570

(*Source:* from Assmann, 1970, p. 79)

2. Natural woodlands or degenerate managed woodlands are unlikely to have the necessary stem volume measurements and thus repeated measurements of stem volume will be required over a five-year period. A number of trees will be clear felled in order that volume/dry weight ratios can be established. This approach then

assumes an accuracy level and a methodology similar to that described for the previous method.

3. Where time does not permit a measurement of volume growth over a five-year or longer period, then accuracy declines rapidly. If it is necessary to complete the production estimate in just one year it is possible to work from the width of annual rings. From a number of felled trees (at least 20 and preferably 35 trees) the radial increments for the previous five to ten years can be measured and a regression of dry weight on volume and of dry matter production on D.B.H. or top height can be applied to the rest of the study area.
4. The least accurate of the methods for estimation of stem biomass change can be made when neither felling of trees is possible nor volume measurements have been made by foresters. The only possibility in these circumstances is to take a succession of increment cores, and to make an estimate of volume increment over a five- to ten-year period. The core would also allow the specific gravity of heartwood, sapwood and bark to be calculated. The increment core method and production estimate from annual rings can, of course, only be of use in seasonal forests. In tropical rain forests where non-seasonal conditions prevail there are no annual rings; this is all the more unfortunate in that rain forest areas are those which rarely have been systematically measured and their extreme complexity with numerous layers of vegetation makes any research into growth rates an exceedingly lengthy and difficult process.

Many of the problems associated with tree growth rates are essentially of a practical nature; for example, which trees should be sampled and how can the sample be identified? Or, which is the most accurate method of measurement? Only experience can tell the project operator that the D.B.H. must be measured at 1·3 m above ground level and on the uphill side of the tree if the terrain is not flat or that a steel tape must be used for all diameter measurements as a linen tape is notoriously liable to stretch in the humid atmosphere of the forest. It is also necessary to produce a large-scale map on which can be shown the sample species, the size classes, dates of sampling and also any other relevant data such as location of meteorological instruments or positions of soil pits.

Yet another problem with both practical and theoretical difficulties is the conversion of volume data to dry weight. Reference has been made to the need for converting figures for biomass change, in terms of volume, to dry weight using the relevant specific gravity values (Table 4.1). Assmann (1970) has discussed this topic and it is useful to summarise his findings. Not only are there the well-known specific gravity differences between the heavier heartwood and the lighter sapwood within a tree but there are also fluctuations along and across the axis of the stem. There are also fluctuations between one tree and the next and between individual stands. Little wonder, therefore, that Newbould (1970) considers the conversion of volume data to dry weight data a low accuracy transaction! Assmann (1970) also suggests that density of wood decreases with increasing elevation above sea level and with higher geographical latitudes.

Assuming that some forest trees have been felled, then a number of measurements must be made upon the freshly felled material: (i) top height; (ii) D.B.H.; (iii) diameter of stem below the lowest living branch; (iv) depth of crown; (v) diameter of crown (best made immediately before felling); and the total fresh weights of (vi) trunk; (vii) leaves; (viii) branches.

Again it will simplify matters if these eight parts are regrouped into: (i) leaves; (ii) branches; (iii) trunks; and (iv) the roots. The trunk should be severed from the roots, at ground level, and all branches carefully removed and put aside. Next, the trunk is sawn into regular sized lengths, 1–3 m long depending upon diameter of the trunk. Samples may be taken from these lengths for calculating density values while incremental growth can also be checked from the annual ring pattern. Results should be recorded either in units of g/m^3 or kg/m^3.

The branches are next assessed. If a regression equation as described on p. 53 has already been established on the basis of the standing trees, then it is both an easy and accurate task to calculate branch dry weight based upon the diameter of the felled trunk. When a regression equation is not available or when branch size and shape is highly variable, then each branch must be weighed separately. An estimate of branch age is also necessary though this may be difficult owing to closeness of annual rings in branches. Next, one must develop a regression equation using age of branch ($\log x$) as the independent variable and branch weight ($\log y$) as the dependent variable and a and b are the regression constant and coefficient respectively. The regression will normally follow a curve described by the equation

$$\log y = a + b \log x$$

When the stem and branch biomass and annual stem wood production are known, then an estimate of branch wood production is possible from the equation

$$\Delta B/B = k(\Delta S/S)$$

Where B and ΔB are branch weight and production, S and ΔS are respectively stem weight and production and k is a secondary ratio of the relative growth rates and may range from 1·86 to 2·00.

Leafy tissue can be assessed in one of two ways. First, it can be considered part of the branch component, or second it may be classified as litter fall. Leaf production is best measured by one of the following methods:

1. the difference between the maximum dry weight of foliage present on the tree during the growing season and the minimum quantity;
2. the dry weight of the tree leaves at the end of the growing season.

The second of these methods is considered by Newbould (1970) to give a more representative result particularly if combined with regular measurements of litter fall. The leaves are weighed and a dry weight conversion obtained from a representative sample. A regression equation is again calculated for dry weight of leaf against the basal diameter of the branch. This will allow an estimate of the leaf biomass for the sample area.

Finally, the biomass of the root component must be assessed. Lieth (1968) has provided a detailed methodology which gives a reliable indication of biomass but requires extensive field work. Depending upon the field situation, all roots over a selected diameter (e.g. 10 mm) are dug out, washed and weighed. Fine roots and those below the selected diameter can be collected from a soil monolith or soil core. Many of the finer roots will have originated from the ground flora and not specifically from trees and thus the fine root fraction must be credited to the biomass of the total vegetation. No simple method exists for conversion of root biomass to root productivity. Newbould (1970) suggests the main problem is that of estimating the turnover of fine roots; these rootlets are constantly being eaten and decomposed by the soil fauna and fungi.

If all four components of the tree (trunk, branches, foliage and roots) can be measured for biomass and if regressions relating biomass to D.B.H. and/or volume change can be made then it is relatively simple to sum the components in order to provide an estimate of stand production. We can now return to the equation used by Madgwick (1970), viz.:

$$\log W = a + b \log D$$

This regression was first suggested by Kittredge (1944) and has been successfully used by many other workers, notably the Japanese (Kimura, 1963, and Ogino *et al.*, 1964). Once the equation has been established on the basis of the destructive sample the equation can be applied to the remainder of the stand by changing the value for D which is the diameter, breast height, of the particular tree under study. Subsequent research has suggested that a slight improvement in accuracy can be made by substituting D^2H for D where H is height of tree.

4.4 THE DETERMINATION OF PRODUCTION FOR MANAGED FORESTS

The natural forest which evolved over long periods of time free from disturbance by man or his domesticated animals is no longer to be found. Even the tropical rain forests of South America and the Far East are rapidly changing because of the incessant search by the commercial forester for new sources of timber. In the past, the natural forest was simply cut over, the commercially useful timber removed and the useless timber left to decompose *in situ*. If left undisturbed and, provided a source of tree seed remained in the area, a secondary forest developed. Often, however, the impact of deforestation upon the soil was so considerable that forest regrowth could not occur and at best a degenerate scrub growth composed of occasional stunted trees amid a grass-dominated flora was the best that could develop, e.g. the *maquis* of the Mediterranean lands.

Increasingly, as the world demand for timber increases and the number of sources of timber becomes depleted, so deforestation is quickly followed by reafforestation. Sometimes, this is the result of direct government control such as can be found in many of the highly industrialised European and Scandinavian countries although it is increasingly becoming part of government policy in the developing African nations. National control of forests has many advantages, not least being the impact forest exports and/or imports can have upon the total trading balance of a nation.

In Great Britain the timber import bill for 1976 amounted to £2 000 $\times 10^6$. Alongside the government-controlled or state forests can be found private afforestation projects. The motives of the private forester can vary from the exceedingly rich land owner who minimises income tax and death duty by investing in forestry to the commercial forestry company which can make a profit out of growing and selling timber. There is little chance of a 'get-rich-quick' policy for the forester; even in low latitude areas where growth conditions are most favourable, it is unlikely that there will be less than 25–30 years between planting and harvesting date. More common is the situation where a forester plants saplings for his successor to fell.

Modern, commercial forestry, be it state-controlled or privately developed, has but one concern; to grow trees for commercial gain as successfully as possible. To this end the diversely specied, varied-aged natural forest is the very antithesis of what the commercial forester requires. His requirement is for straight, even-grained timber of a species which is acceptable by timber merchants for a wide range of purposes. He also must know how much timber exists within his forest, by how much the timber reserve increases each year and when it will be most beneficial to remove trees from the forest.

As a consequence of these demands the commercial forester is not concerned with total biomass or total forest production values. He is, instead, concerned with management techniques which can optimise the production of stem growth and accordingly his concern with forest production methodology is directed to the measurement of often quite small increases in stem timber. For this purpose yield class tables have been devised. The yield class system was devised, as far as can be accurately assessed by Oberförster J.Ch. Paulsen from the Lippe Province of Germany about 1795 but it has only been in the last 100 years that the yield class system has been refined and substantiated by forest research centres which have collected vast quantities of statistical data. Whereas the early yield studies were considered to be minor aids for forest mensuration today the situation is reversed and yield studies provide methods for the assessment of timber increment on which the whole concept of economic forestry rests.

The yield class system has been extensively used by the British Forestry Commission as the means for quantifying growth of the even aged coniferous stands now so common-place in Britain. Forestry Commission Handbooks Nos. 16 (1966) and 34 metric revision (1971) describe in detail the theory and methodology involved in the yield class system. The method has as its basis the actual or the potential maximum Mean Annual volume Increment (M.A.I.) of a measured number of trees in a stand. This can be calculated by dividing the total volume production by the age of the stand. Figure 4.2

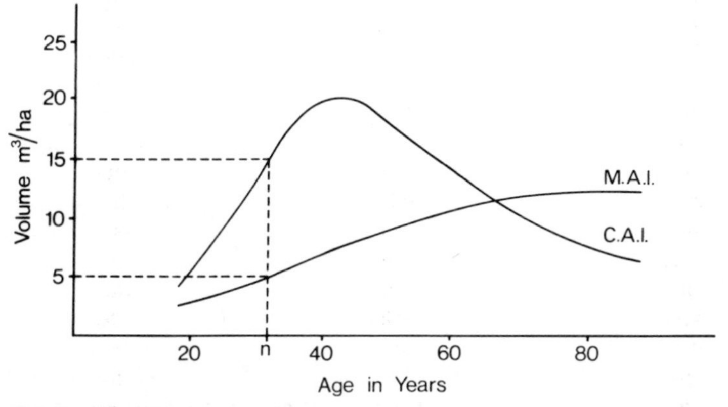

Fig. 4.2 Relationship of mean annual increment (M.A.I.) with current annual increment (C.A.I.) (*Source:* from Forestry Commission, 1971, p. 2)

shows the relationship of M.A.I. with C.A.I. (Current Annual Increment). The latter value shows the typical progressive increase in growth rate until a peak is attained at about 40 years of age. In the figure at age n (32 yrs) the stand presents a C.A.I. volume increment of 15 m^3/ha, while the M.A.I. is the average annual volume increment from the time of planting to n, which in the example is but 5 m^3/ha. The M.A.I. curve attains its maximum value where it crosses the C.A.I., thereafter it falls away. All even-aged stands reveal a pattern similar to that displayed in Fig. 4.2 though the growth potential of a site can change the detailed positions of the curves. Thus Fig. 4.3 shows that the faster a stand grows, the higher the M.A.I. curves will be while the peak M.A.I. value will also occur earlier. The maximum mean annual increment is the maximum average rate of volume production which can be attained by a given species at a site, irrespective of the time of culmination, and as such forms the basis of the yield class system as used by the British Forestry Commission.

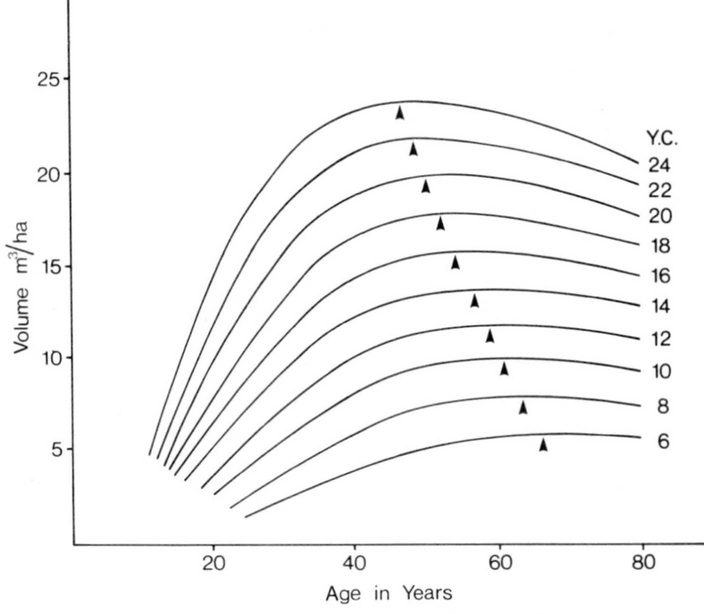

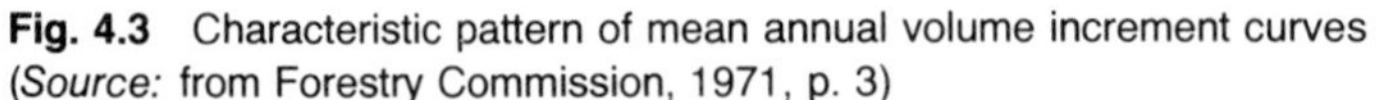
Fig. 4.3 Characteristic pattern of mean annual volume increment curves
(*Source:* from Forestry Commission, 1971, p. 3)

It is usual to refer to Yield Class data simply by quoting a number, sometimes prefixed by YC, thus YC 16, or simply 16, implies that the stand, irrespective of species type, has the potential to grow at a maximum mean annual increment of about 16 cubic metres per annum. Yield class can be calculated in other ways besides using maximum mean annual volume increment. The German approach to yield class calculation has been via the use of tree height/tree age relationships (Assmann, 1970), though the same author has clearly stated that yield class values devised on such a basis will be accurate only if the following two conditions are met.

1. The height growth of the stands must correspond to that of the yield table.

This condition must be checked for the stand under consideration in the field by comparing the height growth of (about) the 40 tallest trees whose height growth must then be compared with mean height development of the stand as a whole.

2. The total crop yield assumed for trees of certain heights must have actually occurred in the stand. Assmann (1970) states this can only be checked for long-term experimental plots.

The British Forestry Commission had previously used the tree height/tree age equation in their early work on timber yield but, partly due to a lack of long-term experimental plots, the accuracy of the yield tables was not universally high. A more accurate description of the tables would be Quality Class tables; the Forestry Commission categorised the stands into one of five quality classes, QC I, II, . . .V (Forestry Commission, 1953). The QC values were specific to individual species types, thus QC I stands of *Larix leptolepsis* would not contain the same volume of timber as QC I *Picea abies*, indeed it would equate to QC II *P. abies*. The Quality Class method of assessing timber production had little value to forest management practice and as such

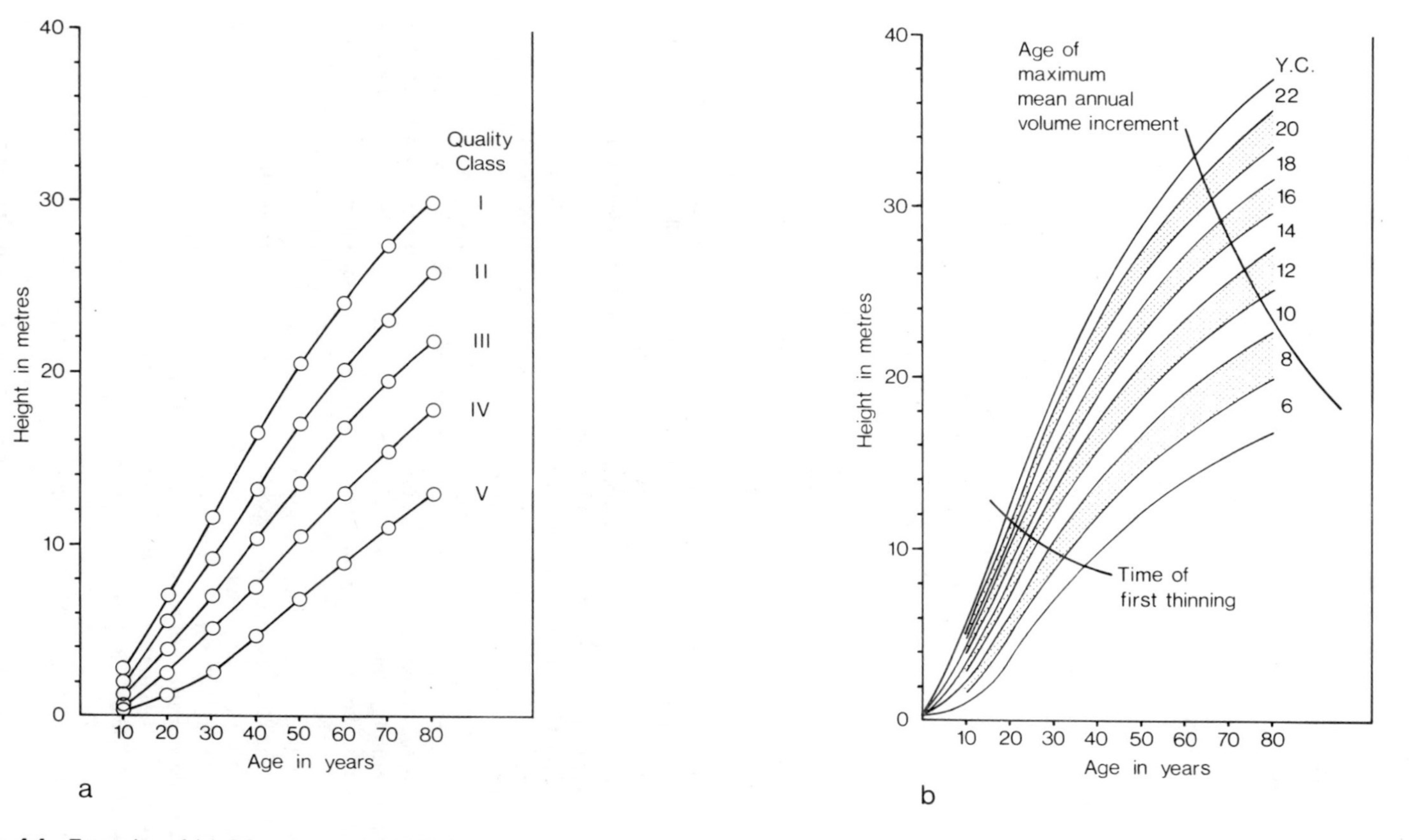

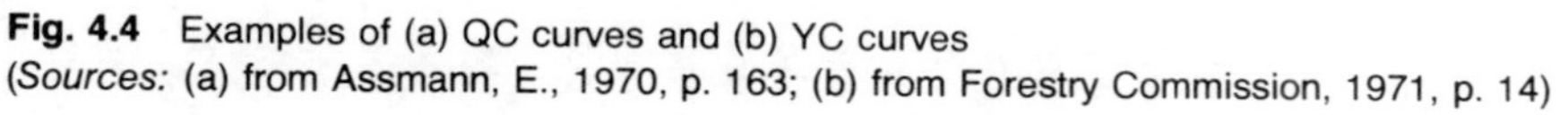

Fig. 4.4 Examples of (a) QC curves and (b) YC curves
(*Sources:* (a) from Assmann, E., 1970, p. 163; (b) from Forestry Commission, 1971, p. 14)

was replaced in 1966 by the Yield Class system described above. As a starting point Yield Class was based upon a tree height/age relationship as was Quality Class, thus YC and QC are comparable, though only indirectly so, for the tree height/age equation was then modified to take account of maximum mean annual volume increment. This statistic was used because it was a measure of the growth potential which was not complicated by the difficult choice of finding a tree age unrelated to the growth characteristics of the species. Volume measurement was also considered superior to height measurement of trees because it had direct parallels for biomass and production values for natural woodlands (see section 4.3). Finally, volume measurement was also known to be a suitable indicator of relative profitability. Examples of both QC curves and YC curves are given in Fig. 4.4 (a) and (b).

Thus, current British forestry practice has strong similarities to the methodology used in the measurement of natural forest production as described in section 4.3. The main difference results from the commercial foresters' prime interest in the stem of the tree and the consequent lack of interest in the production of branches, needles and/or roots. At first sight this may appear a logical decision but if the arguments postulated by Madgwick (1970) are correct, then the foliage and branch production could provide valuable indicators to the overall management level of the forest. No part of the tree should be considered separate from the rest. A compact underdeveloped root system or an uneven crown may be indicators of incorrect planting densities and as such mean production values for all parts of the tree, that is, foliage and branch production, stem production and roots should be assessed and used by foresters in their management strategy. Instead, the Forestry Commission use the Yield Class tables in their existing form to determine both immediate and long term planning. For example, production forecast tables have been compiled; these comprise two separate sections, one concerned with forecasting Thinning Yield and the other with Felling Yield. For both tables the volume yields are given in terms of three different top-diameter sizes, 7, 18 and 24 mm over bark. The figures included in both thinning and felling yield tables have been reduced by 15 per cent to take account of unplanted areas such as roadways, firebreaks, etc. Other tables have been provided to enable the forester to calculate the optimum age at which to thin a plantation or to extrapolate the height at age of maximum mean annual increment. This latter table is of great value for trees which have been planted on soils with a peaty top or on pure peat sites, for on these sites trees which exceed approximately 13 m in height become increasingly prone to wind blow. Using the height at age of max. mean annual increment table, the forester can calculate the lost production through felling too soon. This is an area where the forester has to make a major gamble. Should he allow the trees to attain the maximum increment point and by so doing risk the trees being destroyed by wind blow, or should he fell the trees prematurely thereby forfeiting valuable timber production?

The Yield Class system for classifying production of forests in Britain has been achieved only after the establishment of a number of experimental plots on which parameters of growth such as top height, D.B.H. and biomass have been carefully calculated over periods of forty years or more. Even so the published Yield Class data represent only the latest information from a continuously changing array. Research into forest yield is progressing along different directions in other countries. The highly organised forests of north-west Europe and Scandinavia have become open-air factories, the function of which is to synthesise sunlight, water and basic raw materials into cellulose. Great interest has been shown by foresters in the Dutch research of De Wit *et al.* (1971), who have been experimenting with the modelling of plant growth. Work so far has been applied to agricultural crops and is described as an elementary crop growth simulator (ELCROS). The model assumes that the crop (a single species, or even an entire stand) comprises a reservoir of energy which has been obtained directly from the process of photosynthesis. The plant energy is subsequently used for life and growth processes (Fig. 4.5).

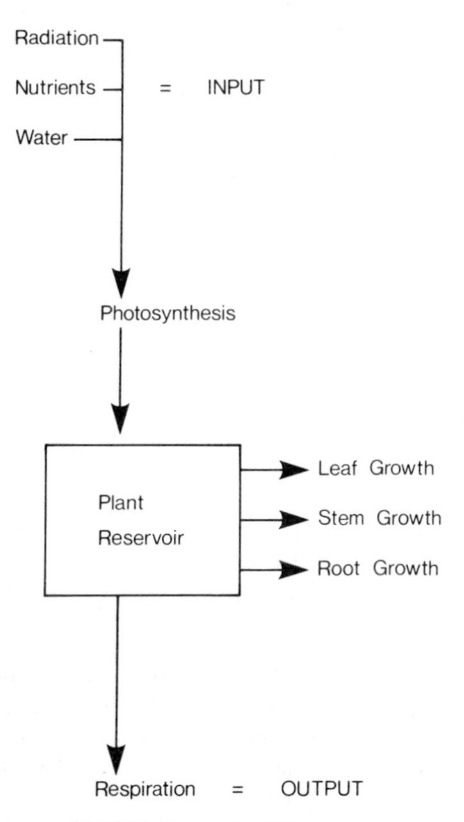

Fig. 4.5 Simplified diagram of ELCROS crop simulation model
(*Source:* from De Wit *et al.,* 1971, p. 118)

Respiration represents a direct output of energy back to the environment while the passage of energy to structural leaf growth, stem growth and root growth all represent a transfer of energy into organic components. It is the rate at which energy is transferred to these organic components which determines the production level of a plant. To enable ELCROS to function all the energy flows have to be identified and translated into a computer-based language. This has been done on a general scale – but has also revealed further gaps in knowledge. The complete simulation of growth processes is, at present, a slow, cumbersome task but results are sufficiently comparable with reality to suggest that growth models will, at some time in the future, be capable of allowing foresters, agriculturalists and biologists to play through various management techniques on a computer before ever moving into field trials. For forestry this has immense implications. The life span of trees is measured in tens of years, traditional field experimentation requires perhaps twenty years before the first results are obtained. Using an ELCROS-type approach, results could be gained in twenty months!

Not all forest policy has moved as rapidly as described in the previous paragraph.

Somewhat surprisingly, forestry in the USA has, according to Avery (1967), been using methodology 'just as foresters did five decades ago'. Such a criticism was certainly not applicable to the pioneering work of Whittaker and Woodwell (1968), who systematically established weight, production and surface relationships for seven species of trees and shrubs in an oak-pine forest on Long Island, New York. Using methodology akin to that described at the beginning of this chapter Whittaker and Woodwell (1968) were able to establish allometric regression equations between measurable growth parameters and biomass and to establish production data which would satisfy both the needs of the forester in that estimates of merchantable timber could be predicted from the regressions while the applied scientist could also obtain data on net plant production for the forest ecosystem as a whole.

It is clear that research on the primary production of forests is still in its early years of development. Methodology exists for N.P.P. values to be calculated for natural, multi-species, uneven-aged stands but the techniques are time-consuming and their accuracy levels are frequently of low value. Accurate forest mensuration of even-aged, single-species stands has been achieved in Britain but the data is available only for the trunk wood with no recognition having been given to roots, branches and foliage. Sufficient data are available to suggest that the world's forests have an N.P.P. (77×10^9 tonnes) which is greater than any other land-based ecosystem, or indeed the entire ocean system. This is sufficient to show the great biological significance of the forest system as a production unit upon this planet.

The importance of mathematical modelling of forest systems holds out great hope for a major breakthrough in understanding forest production. To this end, the work of De Wit *et al.* (1971) marks an important stage in the attainment of this goal. Similarly, the work of Botkin *et al.* (1972) on the construction of a computer model to describe forest growth for a mixed-species forest in the Hubbard Brook Ecosystem Study represents the first successful application of computer simulation to such a complex ecosystem. The advantage of the approach of Botkin *et al.* (1972) over the traditional regression and curve-fitting approach as described by Newbould (1970) is that computer simulation can serve as a repository for all ecosystem information and, of even greater relevance, repeated hypotheses can be tested using Monte Carlo samples of simulator runs allowing comparison of simulated and field observations. Continued research in this area is likely to permit a rapid and considerable progress on the production levels of world forests both for the remnants of natural forest and for the extensive tracts of commercial afforestation.

CHAPTER 5
A CASE STUDY. THE CALCULATION OF SITE PRODUCTION INDEX

5.1 THE CHOICE OF STUDY AREA

The example chosen for a case study was a highly managed Forestry Commission forest in Dyfi Forest, Wales (Fig. 5.1). It may be considered a somewhat unusual choice for a case study but the forest type was deliberately chosen for a number of reasons. Natural vegetation is now confined to small reserve areas and is of such high value, and often of such limited extent, that productivity work is not possible. The place of virgin vegetation has been taken by the semi-natural or anthropogenic-controlled vegetation, plant systems in which man directly or indirectly controls plant growth. The case study presented in this chapter is, therefore, a realistic study of vegetation production from a highly managed vegetation system in which all stages of growth are supervised by man. Managed forest systems are readily accessible over much of north-west Europe, North America and also in Australia and New Zealand and the methodology described in this chapter should allow students to repeat the work at local sites. Managed forests, in contrast to natural forests, are generally very much simplified ecosystems in that species variation, species age, vertical structure and ground flora show minimal variation over a small area. This allows 'representative' sites to be chosen with comparative ease. In most instances the managed forest will have been monitored for growth rate throughout its life by professional forestors. The timber volume per hectare will also be known. Consequently, production data can be more readily calculated and instead of a five-year study necessary for natural forest, the managed forest can yield results within a year. But there are some disadvantages to this artificial forest type. It has a monotonous regularity; its production level can be the result of particularly good, or bad, management policy; the production values give no indication of total biomass and are often for exotic tree species. These features must be considered when interpreting results. It can be argued that total biomass values are really an extravagant luxury for if man cannot use the leafy material along with branches and roots there is little point in measuring them. As the previous chapter has shown, their measurement demands very considerable time and the results can be of dubious accuracy. Total biomass is, however, the only real indicator which can provide information on the total relationship of the plant with its environment.

The case study presented here attempts to show how one tree species, Sitka spruce (*Picea sitchensis*), when planted at different elevations and on different soils, can produce variable growth performance. This species was chosen as it represents the most important tree, in an economic sense, throughout Great Britain.

Discussion with the District Forest Office at Dyfi Forest revealed a suitable area for the calculation of Site Production Index (S.P.I.). This was the Cwm Cadian 'beat' (district), an area of a little over 1 000 ha, located on a south-south-east facing valley side (Fig. 5.1). Geology was considered to be uniform, consisting of Silurian and Ordovician silts and shales with varying amounts of till, scree and head material producing soils of very differing depths (from 50 to 1 500 mm). Elevation of the site ranged from 65 m to 550 m O.D. With such a wide range of elevation it was considered probable that micro- and meso-climatic variation would be responsible for considerable variation in timber production.

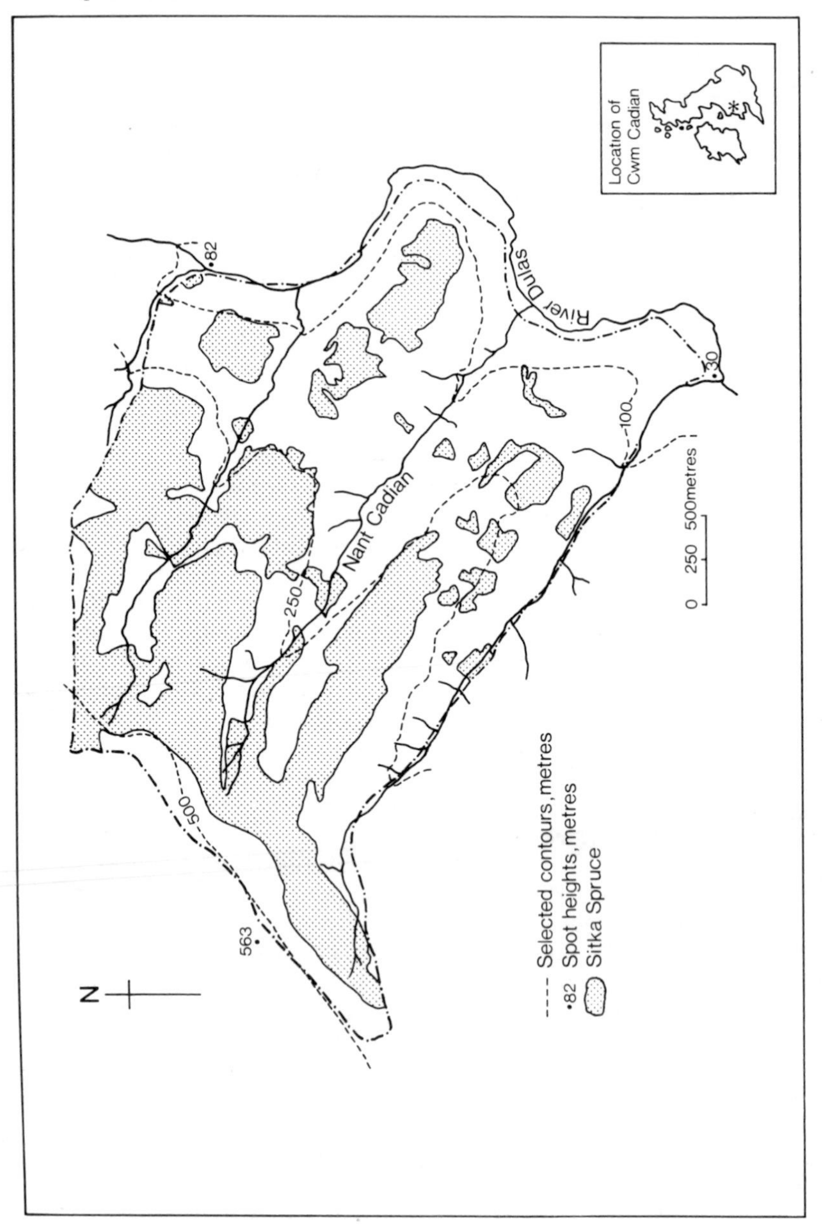

Fig. 5.1 Dyfi forest, location of study area. (Loosely based on an Ordnance Survey Map. Crown Copyright Reserved)

5.2 SITE CONSIDERATIONS

The Forestry Commission had subdivided the area into 'compartments'; these were areas defined on a map by a roadway, ride, fire break, forest margin or tree species difference. The original compartment pattern shown in Fig. 5.2a had been set up in the 1920s when planting first began in the area and was gradually extended as the forest

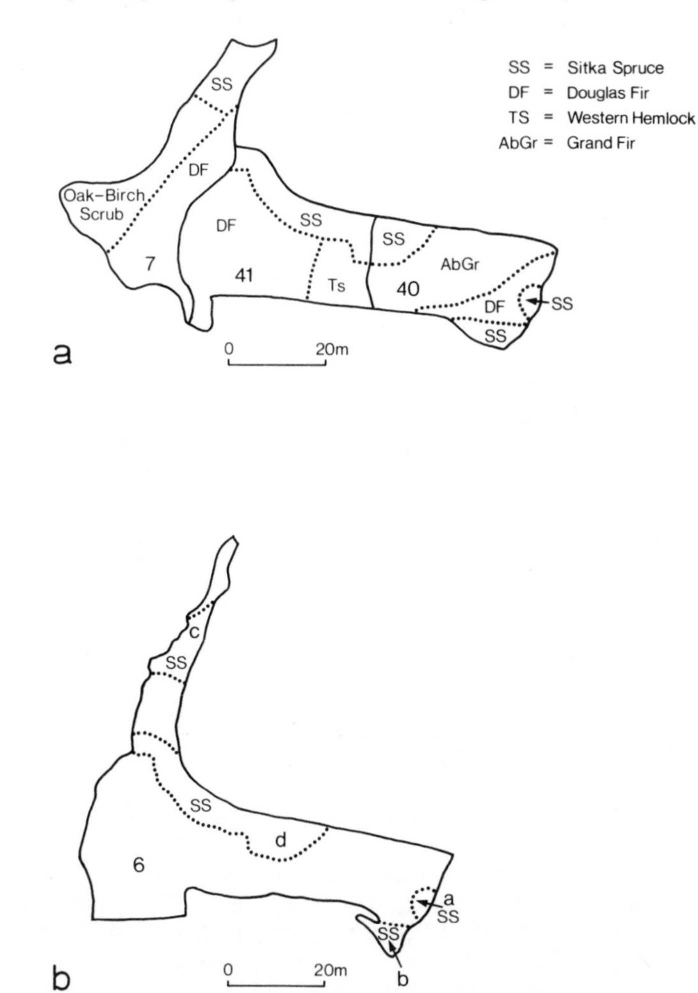

Fig. 5.2 Dyfi forest Compartment pattern. (a) Pre-1968, (b) Post-1968
(*Source:* based on Forestry Commission working maps)

increased in extent. Only rarely did each compartment contain a single tree species type throughout; more frequently the compartment had been divided into sub-compartments with each sub-compartment containing (usually) only one tree type. The trees in each sub-compartment had been subjected to detailed measurements and these are shown in Table 5.1a.

The chief indication of suitability of the species to a site was given by the QC value (Quality Class). As was indicated in section 4.4, the QC value gave no indication of the

Table 5.1a Forest management details for Compartment 41, Cwm Cadian, Dyfi Forest. Pre-1968 data

STAND NO.	PLANTING YEAR	SPECIES	TOP HT. (feet)	QC	PER ACRE		AREA 24·6 acres	
					B.A.	Vol.	Stand area	Vol.
a	28	Ts	65	I	90	3 000	6	18 000
b	28	DF	55	III	65	1 650	12	19 800
c	39	JL	37	IV	70	1 200	2	2 400
d	55	SS	3–4		*not assessed*		4	

(*Source:* from Forestry Commission working documents)

Table 5.1b Forest management details for revised Compartment 6, Cwm Cadian, Dyfi Forest. 1968 data

Sitka Spruce only considered

STAND NO.	PLANTING YEAR	TOP HT. (feet)	YC	PER ACRE TOTAL PRODUCTION		STAND AREA	TOTAL VOL.
				B.A.	Vol.		
a	28	56	200	278	7 515	1	7 515
b	55	6–9		*not assessed*		2	
c	27	18	180	261	6 615	2	13 230
d	55	6–9		*not assessed*		4	

Note: Data is still in Imperial units of measurement. Further changes occurred in 1971 when all forest measurements were converted to metric units.

(*Source:* from Forestry Commission working documents)

expected growth rate and as the forest grew to maturity the inadequate nature of the QC value became greater. Accordingly, the Forestry Commission devised the General Yield Class (GYC) method of assessing tree performance and this replaced the QC through the 1960s. The change necessitated a re-defining and a re-drawing of compartment boundaries. From 1968 compartments included stands of similar GYC values though not necessarily of the same species type for GYC values, unlike the QC values, were comparable between different species. Thus, all compartments of like GYC values could have similar management techniques applied to them. An example of the revised compartment and data format is given in Fig. 5.2b and Table 5.1b. By means of a conversion factor pre-1968 data could be made compatible with the new data and this process allowed comparison of two independent production assessments made at different stages in the life history of the forest.

The attainment of potential production from a site is a delicate combination of individualistic management and predictive skill in forecasting future demand for specific timbers. This process begins with the preparation of the site for planting followed by the correct choice of tree species with precise matching of provenance to the prevailing environmental conditions at the planting location. Site preparation will vary depending upon terrain and may involve only shallow ploughing to disturb the existing vegetation and topmost soil horizon. Conversely, difficult sites may require deep ploughing to disturb pan formations 500–600 mm below the surface while deep peat may require frequent and large drains in an attempt to reduce the water table and in turn improve potential root growth (Forestry Commission, 1963). Correct site preparation is critical for the future wellbeing of the stand. Failure to thoroughly disturb any iron pan or hard pan in the soil profile could lead at worst to a cessation of growth by the developing trees or at least to a period of 'check', a minimal growth period for upwards of 15 or 20 years, until the trees grew though the impediment (Forestry Commission, 1954). An underdrained site can have disastrous consequences upon tree growth. An excess of water in the soil profile inevitably means a deficit of soil air, possibly resulting in seasonal or permanent anaerobic conditions. Under these conditions root development is impaired and a satisfactory anchorage is impossible to achieve. It has been shown that conifers planted on poorly drained, peaty sites have a greatly increased chance of being uprooted, or windblown, when they exceed 12 m in height. Conversely, overdraining a site, or planting the incorrect species on a freely drained site can lead to pronounced soil water deficits during the late summer months which can reduce growth rate and impair the quality of the timber making it too brittle and prone to stem-snap. Site preparation is a time consuming and expensive procedure; defective preparation gives the forester a legacy of problems for the remainder of the life of the forest.

The choice of tree species and seed provenance are two major decisions that the forester has to make early in the process of afforestation of a site. In the past, choice of species was determined largely by site factors, e.g. steep valley slopes were given to *Larix* spp. (Larch) as they were considered best able to tolerate the free drained sites while frost hollows were devoted to *Picea abies* (Norway spruce) as they were believed to be frost hardy. Since 1970, two factors have changed this policy. First, foresters are increasingly influenced by the predicted market potential for timber in 50–100 years hence. It is forecast that the major demand will be for a timber suitable for the manufacture of pulp and timber quality becomes of secondary importance. Provided that the trunk is straight for approximately 2 m it will be acceptable by the pulp mill. Maximum production with minimum concern for quality is thus the aim and to this end *Picea sitchensis* (Sitka spruce) is pre-eminent. The second factor responsible for the change in planting policy has come about through the forester's ability to manipulate the site in order that it is suitable for a specific species. Thus, increasingly, land to be afforested is prepared for *Picea sitchensis*. Great advances have been made in understanding the site requirements of this species particularly in relation to the 'provenance' of the seed. The provenance refers to the origin of the seed-bearing parent trees. Until the late 1950s it was thought that it was necessary to know only the regional location of the seed source but more recent work has shown that precise matching of seed source with the ultimate planting area is crucial to really successful afforestation projects. The success of *Picea sitchensis* as a plantation tree within the British Isles is, in no small part, due to its native distribution pattern in the north-western area of North America. Not only is the altitudinal range at which this species can grow particularly wide but there are also distinct sub-types to be found at low elevation coastal and low elevation continental sites as well as mountain sub-types. It is now known that seed from particular areas on Queen Charlotte Isle, British Columbia, grow best in the maritime conditions to be found along the western seaboard of the British Isles.

Provided the highly important site preparation and species choice have been correctly made, the young transplants of about three or four years old will grow satisfactorily in their new habitat. On nutrient deficient soils a slow-release fertilizer application may be made during the early years of life in order to boost tree growth rate. It is customary for early growth to be erratic between years. No production measurements should be attempted on an immature forest – that is younger than 30 years of age, for during this initial period the growth rate may fluctuate from as little as 50 mm to as much as 600 mm per annum. Thinning will also occur at least once during this initial period, for *Picea sitchensis* the date of first thinning would commence at 18 years after planting at the better sites and at the 32nd year on the poorest sites. The objective of thinning is to maintain tree density at such a level that growth will be vertical, so producing the desired straight main stem, but at no time must tree density be such that growth rate declines because of a lack of growing space, both for branches and

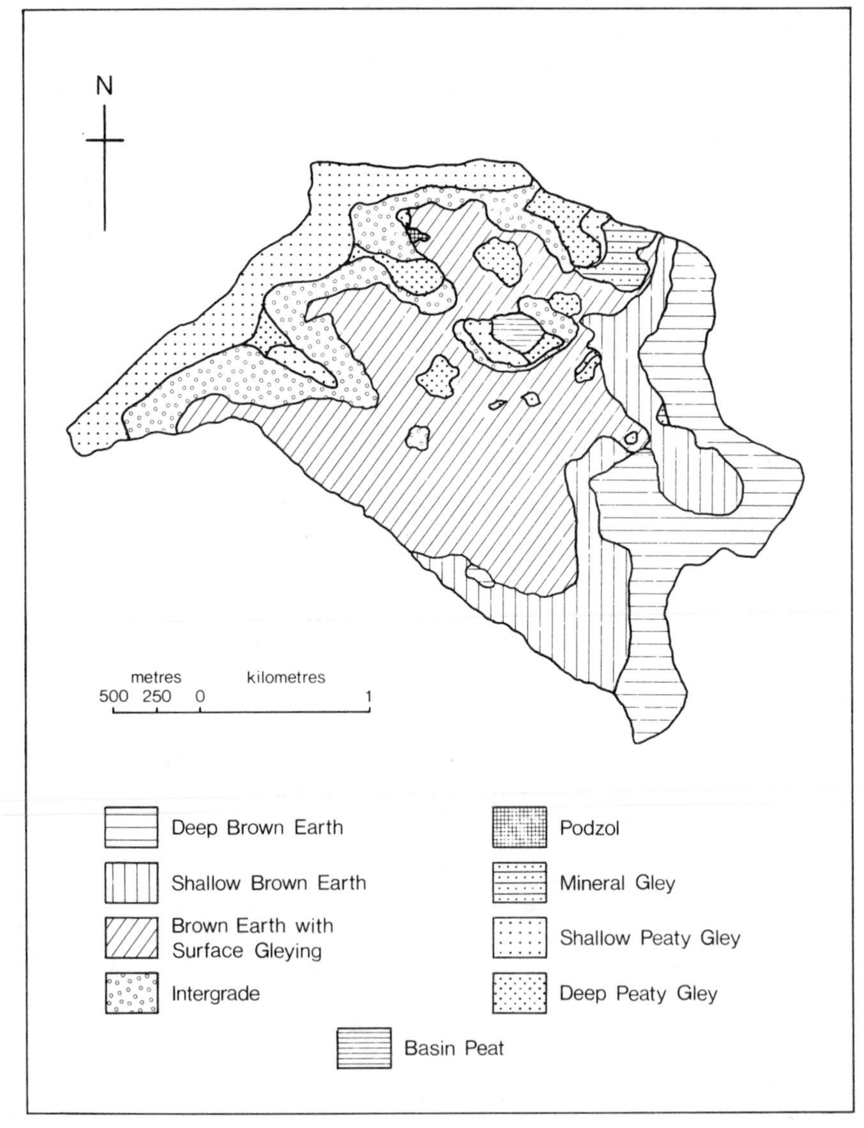

Fig. 5.3 Soil map of Cwm Cadian

for the below ground root structure. To the forester, thinning is entirely a management practice to improve the eventual timber production level of the stand. To the biogeographer, thinning is a nuisance for it represents a major change in the structure and biomass value of the stand. For example, *Picea sitchensis* of GYC 18 would have an approximate density of 3 237 trees/ha before first thinning. After thinning, at about 20 years, density would fall to about 2 000 trees/ha while at stand age 30 years density would be about 900 trees/ha. Thereafter, tree numbers decline much more slowly until at 80 years the stand would contain just over 253 trees. The effect of thinning upon environmental conditions within the stand are dramatic. A thinned stand appears light and airy by comparison to an unthinned stand. A ground flora will become established, albeit temporarily, until the canopy grows into a closed condition, i.e. until the needles of adjacent trees begin to touch.

An unmanaged forest, by contrast, be it deciduous or evergreen, hardwood or softwood, will not be subjected to the periodic upheavals of timber removal. The natural forest, being composed of uneven aged, multispecies trees, will exhibit a continuing saga of life, death and decay. At no time is there likely to be a substantial loss in biomass. The only exceptions would occur after fire had ravaged the forest or following a violent storm during which time trees would be blown down. In the latter case the fallen trees would be on the forest floor and would still contribute to total biomass.

The managed forest achieves only the briefest periods of constancy, these being confined to the occasions between thinning operations. It is during these periods and provided the forest is older than 30 years that forest production data may be gathered. The Dyfi Forest case study met both these assumptions. There were, however, some other problems. The range of elevation already referred to would create very different environmental conditions at the lowest point in the forest compared to the highest. The conditions would involve not only direct-acting forces such as temperature, precipitation and exposure values but also a host of indirect and often unmeasurable factors such as evapo-transpiration ratios and the proportion of time a soil remained waterlogged and/or aerated. As a result of the operation of these factors the growth potential at Cwm Cadian would be more favourable on the lower slopes where protection from wind-exposure, low temperatures and high rainfall would combine with deeper soils and appreciable flushing-in of nutrients by lateral moisture movement to produce sites of considerable growth potential. Conversely, the upper reaches of Cwm Cadian would be physically exposed sites, with little protection from climatic extremes and thin, impoverished 'shedding' soils. These sites would be expected to show the poorest growth potential. Figure 5.3 shows the main soil types to be found in Cwm Cadian.

5.3 THE CALCULATION OF THE PRODUCTION INDEX

The method used for calculating the site production index is set out below.

1. The distinct, single species stands of *Picea sitchensis* of 30 years of age or older were mapped, indexed and tabulated.
2. The soil type boundaries occurring within the *Picea sitchensis* stands were plotted on a large-scale map, 1 : 10 560, and the soil types which occurred in each compartment were indexed and tabulated alongside the data from 1.
3. The extent of the different soil types in each stand of *Picea sitchensis* were calculated and listed.

4. The volume in m^3/ha per stand as obtained from Forestry Commission yield class data was listed.
5. The extent of each stand of *Picea sitchensis* in each soil type was listed.
6. The average volume per ha per compartment per annum (i.e. the volume per hectare per stand divided by the age of the stand) was then calculated.
7. As the Forestry Commission data was based upon the compartment, irrespective of soil boundaries, the average volume per ha per compartment per annum obtained in 6, above, was applied to each soil type occurring within the boundary of each distinct compartment.
8. The individual average volume per ha per compartment per year per soil type obtained in 7, above, were summed. The result was divided by n (where n was equal to the number of stands or parts of stands occurring over each soil type), to give the average volume production per soil type.

This latter value was used as the Site Production Index (S.P.I.). Because of the procedure used in steps 1–8, the S.P.I. values calculated for each compartment of *Picea sitchensis* were specific to particular soil types and as the soil types themselves were located at distinct elevation bands then S.P.I. values could also be related to specific elevation groupings. Table 5.2 shows the site production indices for 8 (1955 data) and 7 (1968 data) soil types respectively.

Table 5.2 Site production index for Cwm Cadian, Dyfi Forest. Index based upon rate of growth of Sitka spruce

SOIL TYPE	PRODUCTION INDEX		PERCENTAGE INDEX	
	1955*	1968†	1955	1968
Deep Brown Earth	83·1	220·5	100·0	100·0
Shallow Brown Earth	78·1	175·2	93·9	79·4
Brown Earth with surface gleying	59·3	104·1	71·3	47·2
Intergrade	46·8	81·0	56·3	36·7
Ironpan soil		*insufficient data*		
Mineral gley	*no data*	149·4	*no data*	67·7
Shallow peaty gley	42·4	82·0	50·9	37·1
Deep peaty gley	54·6	101·0	65·7	45·8
Basin peat	38·0	*no data*	45·7	*no data*

*Values expressed as Hoppus feet increment per acre per annum
†Values expressed as average annual increment, Hoppus feet per acre

5.4 RESULTS OF PRODUCTION INDEX

From Table 5.2 it is apparent that maximum tree production could be found on the Deep Brown Earth soils. These soils were to be found on the often steep, lower valley sides, between 50 and 100 m O.D., and had wherever possible been retained by local farmers as valuable grazing land. The inherent value of these sites is substantially due to their location. The lower valley sides are protected from the climatic extremes which occur higher up slope, though cold air drainage can be a problem particularly in springtime and autumn. Also the location permits the maximum input of flushed nutrients thus maintaining soil fertility. Based upon the 1968 data, which was considered to be the more accurate of the two mensuration surveys, the Deep Brown Earth sites were capable of 20 per cent more production in terms of stem wood from *Picea sitchensis* than any other part of the study area. As an indication of suitability of these soils to support tree growth one stand of Western Hemlock (*Tsuga heterophylla*) planted in 1928 on Deep Brown Earth soils in Cwm Cadian had grown so rapidly that it had become recognised as one of the best stands of the species in the whole of Europe.

Where the depth of the Brown Earth soil became less than 450 mm the soil was described as a Shallow Brown Earth. These sites extended from 75 m to 200 m O.D. and the increased elevation caused a deterioration of the physical environment so much so that a decline in actual timber production values also occurred. Even though the inherent production values of these sites was less than for the Deep Brown Earth sites it was still considered worthwhile by the local farmers to retain much of this soil type for sheltered grazing land. At Cwm Cadian only three stands of Sitka Spruce older than 30 years could be found on the Deep Brown Earth, while 13 stands occurred on Shallow Brown Earth which suggested that for many farmers the profitability and the increasingly difficult nature of the terrain at the latter sites justified sale of some of the land to the Forestry Commission. A further increase in elevation to the range 150–365 m O.D. brought about a marked decline in potential production values so much so that agricultural competition was eliminated and forestry had exclusive use of the land. The soil had deteriorated further; increased wetness and decreased evaporation resulting from a wetter and cooler climatic input resulted in seasonal waterlogging of the surface layers of the soil profile which in turn had produced evidence of gleying. Production of timber was now less than half that of the lowest elevation sites. The decline in timber production was exceedingly rapid over the range 50–365 m, thereafter further increases in elevation had a much smaller effect upon timber yield. Wherever drainage conditions became impeded – most frequently because of local depressions in the landscape – then soils became heavily gleyed. At elevations below 365 m the resultant soil was classified as a Mineral Gley as opposed to a true Gley-type soil for there was considerable profile evidence to suggest that gleying was a seasonal phenomenon for the Ag and Eg horizons comprised strong grey-brown mottling of the loam. The basin-like topography made such sites local reception areas for flush waters and in theory should prove to be of high inherent fertility. Of the few stands of *Picea sitchensis* which had been planted on Mineral Gley soils all had shown surprisingly good growth, showing a 20 per cent increase on production levels when compared with the Brown Earth with Surface Gleying. Management of the Mineral Gley sites was difficult as they proved difficult to drain adequately; their superior production potential, however, was reward for these efforts. These sites exhibited a greater degree of shelter than did the adjacent Brown Earths with Surface Gleying. This shelter value may therefore be assumed to contribute to a proportion of the improved production values.

The sequence of production zones as related to soil type becomes confused above about 365 m O.D. The next soil type to emerge was the Intergrade, a soil of somewhat dubious evolutionary processes, for the profile characteristics showed similarities to both the Brown Earth with Surface Gleying and to the Iron Pan Podzol. Intergrade soils could most easily be identified from the gound vegetation and the topographical location. Invariably they occupied convex slopes between 300 and 450 m O.D. and were covered by species such as *Calluna vulgaris, Vaccinium myrtillus* and *Ulex galli.* The rate of production of *Picea sitchensis* was lower on this soil type than on any other at

Cwm Cadian for which data was available. Possible reasons for the poor production rates might be found in the shallow profile depth – just 500 mm of rooting depth, though contained within this was an iron pan which restricted root movement. The typical location of these soils would also create difficult growth conditions, convex slopes with high exposure values and with very rapid drainage characteristics which would produce adverse conditions for much of the growing season.

Occupying a similar elevation zone as the Intergrade soils but in very different topographic situations were the Deep Peaty Gleys. Wherever drainage became impeded then an accumulation of peat formed above the mineral soil. When this layer exceeded 450 mm the peat was classified as deep peat. Provided that adequate drainage channels had been cut then the Deep Peaty Gley proved to be surprisingly productive considering the altitude at which it was found (from about 300–420 m O.D.). When peat depth was less than 450 mm the final soil type was encountered, the Shallow Peaty Gley. This soil type was confined to the highest ground with steepest slopes, 460–560 m O.D., and trees planted at this height were located at or near the present tree line.

Although timber production rates from the Intergrade, Deep Peaty Gley and Shallow Peaty Gley soils fell very far short of the rates obtained from soils of the lower valley slopes it is interesting to note that the fastest decline in production potential occurs on these high production sites. Table 5.3 shows the pattern of production decline.

Table 5.3 Percentage production decline for Sitka spruce with increasing elevation. Cwm Cadian, Dyfi Forest. 1968 data

SOIL TYPE	MID-POINT ELEVATION IN METRES	PERCENTAGE PRODUCTION INDEX	PER CENT PRODUCTION DECLINE FOR EVERY METRE INCREASE IN ELEVATION
Deep Brown Earth	75	100	–
Shallow Brown Earth	137	79	0·30
Brown Earth with surface gleying	257	47	0·26
Intergrade	375	37	0·10
Deep peaty gley	363	45	0·009
Shallow peaty gley	512	37	0·06

Site production has been measured in terms of the growth rate of *Picea sitchensis* on specific soil types. It has been assumed that the so-called 'environmental factors' such as exposure, altitudinal differences, angle of slope and aspect had been automatically included in the determination of the site productivity index through the use of (*a*) the total growth rate of *Picea sitchensis* and (*b*) the complete soil profile characteristics. Both the tree and the soil had developed in response to the stimuli obtained from the total environment and, provided that the range of site conditions encountered in the field are not too large, then the influence of these conditions may be considered to exert a relatively uniform effect upon tree growth and soil development.

5.5 RELEVANCE OF SITE PRODUCTION SURVEYS

The Site Production Index has a fundamental practical value to foresters. The aim of the few quantitative site assessment projects which have been conducted in Britain has been to develop a means whereby the potential production of a tree species at a site can be predicted with sufficient accuracy as far into the future as is possible. While it is unlikely that planting decisions would ever be made solely on the recommendation of a quantitative site assessment study, the professional forester desperately requires more information to assist him in selecting, and of equal importance, in predicting the potential timber production output in *n* years hence. At the present the decision to plant species X at site Y is made only after a careful study of the ecological and economic factors which operate at a site. Each tree species can be imagined to possess a series of 'tolerance zones' to specific site inputs and the more accurately these inputs can be matched against the requirements of the tree the better are the chances of optimum site potential production being achieved. Table 5.4 sets out the most important site inputs relevant to tree growth in Britain.

Ford and Fraser (1968) maintain that in Britain, forest production is limited by factors other than solar radiation inputs. Instead, there are a range of environmental conditions many of which can be modified by management; for example in Table 5.4 the inputs of soil type and drainage conditions would fall into this category, for management processes allow manipulation of both these variables. Other inputs are immutable, again reference to Table 5.4 reveals that the topographic shape of a site and all of the climatic inputs are unchangeable. The fact that a large proportion of the inputs are fixed only serves to strengthen the argument that for successful production levels to be attained the forester must recognise all the inputs that are operational at a site and that the magnitude of these variables should also be known. At this point it is relevant to recall the work of De Wit *et al.* (1971) and the attempts to produce a computer simulation for plant growth. Once such a simulation has been produced for controlled forest experiments it is then but a short step to relate growth to revenue. The S.P.I. index for Cwm Cadian has shown that on favourable sites annual production can attain GYC 24 while on the poorest sites GYC values of 6 are obtained. An improved understanding of growth processes and requirements should, at some point in the future, allow an overall lifting of GYC values. The impact of raising a GYC value of 10 to 12 or even 14 would have a significant impact on forest economics.

Computer simulation of forest growth is still in the gestation period and it will be a decade or more before the technique becomes available for field use. In the meantime it should be possible to use the accumulating wealth of production data from the maturing forests and to examine the effects of site inputs upon production level. Several attempts have been made (Page, 1970; Dixon, 1971; Jones, 1972) to regress production values against a number of the more readily measured soil and topographic factors. Page (1970) used multiple regression analysis to examine 150 forest plots in Gwydyr Forest, North Wales. Using *Picea sitchensis* stands growing only on podzolic soils Page produced the following regression values.

$$Y = 119{\cdot}82 - 0{\cdot}042X_1 + 4{\cdot}28X_2 + 3{\cdot}92X_3 + 2{\cdot}62X_4 - 0{\cdot}13X_5 - 59{\cdot}83X_6 + 27{\cdot}04X_7$$

Where Y = top height at 50 years
X_1 = elevation (feet)
X_2 = shape of slope (numerically converted)
X_3 = shape of contours (numerically converted)
X_4 = soil colour at 6 inches (Munsell)
X_5 = per cent moisture at 6 inches (dried at 105°C for 24 hours)
X_6 = bulk density of soil at 6 inches
X_7 = bulk density of soil at 12 inches

Table 5.4 Site inputs to be considered prior to conifer tree planting

Elevation (metres)	<91	91–122	122–152	152–183	183–213	213–244	244–274	274–305	305–335	335–366	366–396	396–427	427–457	>457
Site conditions	Flat	Gentle slope			Steep slope									
		Convex	Linear	Concave	Convex	Linear	Concave							
Average annual precipitation (mm)	<762	762–864	889–990	1 016–1 118	1 143–1 245	1 270–1 372		1 397–1 499	1 524–1 626	1 651–1 753	1 778–1 880	>1 905		
Soil type	River alluvium	Brown earth		Brown earth with surface gleying	Intergrade	Iron pan Podzol	Peaty Podzol	Gley	Peaty-Gley	Peaty-Gley Podzol	Organic	Sand	Industrial waste	
		Deep	Shallow											
'Semi-natural' vegetation	Molinia dominant	Molinia and/or nardus	Nardus and/or molinia	'Soft' grasses	Juncus-sphagnum	Des-champsia dominant	Pteridium dominant	Ericas dominant						
Natural drainage conditions	Free		Good		Moderate		Poor		Impeded					
	Always	Seasonal	Always	Seasonal	Always	Seasonal	Always	Seasonal	Always	Seasonal				
Frost-free period. mths/yr	5	6	7	8	9	10	11	12						
Aspect in degrees	0–45	46–90	91–135	136–180	181–225	226–270	271–315	316–360						
Exposure index value	Blue	Yellow	Green	Red	Purple									
Former land use	Deciduous forest	Coni-ferous forest	Degenerate woodland	Heath	Moorland	Grassland		Agriculture	Sand dunes	Industrial waste				
						<183 m	>183 m							
Special features	Coastal site	Forest adjacent to industry	Forest adjacent to urban area	National park 'preserved' area	Picnic site, nature trail	Frost hollow								

The work of Jones (1972) used the information gathered for the calculation of S.P.I. values at Cwm Cadian (section 5.3) and a multiple correlation for 38 forest plots was made with the S.P.I. values as a dependent variable and independent variables as follows:

X_1 = average elevation of stand in feet
X_2 = age of tree in years
X_3 = average slope of ground
X_4 = aspect index value
X_5 = GYC
X_6 = top height of trees in feet

Variables X_2, X_5 and X_6 were later omitted because it was shown that these variables were highly interdependent. Production levels were then shown to correlate with aspect, average elevation and slope giving a multiple correlation value of $r = 0{\cdot}957$. Very similar results were produced by Dixon (1971) for Douglas Fir (*Pseudotsuga taxifolia*) production in Scotland when topographic position and exposure were particularly relevant in describing production values.

The results of multiple regression and multiple correlation exercises are highly encouraging. The techniques can be profitably used to show the level of explained variation using a number of variables though there are also some qualifications to the use of these methods. Most important, the statistical techniques are only valid if the data on which they are used are independent variables. From work at the Cwm Cadian site (above) it appears that remarkably few of the variables are completely independent. Two groups can be identified; those which may be classified as site parameters such as elevation, annual precipitation, aspect, exposure index and soil characteristics. In this group, elevation is the master factor with all other variables showing some degree of interdependence upon it. The second group includes all the growth measurement variables such as yield class values, top height of tree, age and diameter at breast height. As top height is basic to the calculation of production values it can usually be selected as the master factor. Another major area for concern involves the unknown and unmeasured variables. For example, in the examples quoted in the previous paragraphs the variables were chosen for two reasons; first, they were considered the most relevant variables and second, they were readily measurable either in the field or the laboratory. There may well be variables of much greater relevance which, due to ignorance or lack of measuring ability, have been omitted. With present levels of knowledge we can but assume that the relevant variables have been used. It is probable that with help from simulation modelling of ecosystems that other variables will emerge as being worthy of inclusion in multiple regression and multiple correlation projects.

The demand for quantity of timber is likely to continue to increase throughout the immediate future. Schery (1972) has predicted a ten-fold increase in demand by the year AD 2000, and even the vast forests of northern latitudes and of the tropics will be unable to meet these demands. Intensification of output from the highly managed forests of north-west Europe will become increasingly important and, with the continuing upward spiral of land prices, it is imperative that commercial afforestation makes every attempt to increase its production in order to justify its use of extensive tracts of land – for which alternative uses will also be increasing. This increase in production must be achieved along sound ecological lines otherwise site degradation will become inevitable and once again man will be criticised for sacrificing short-term maximisation against longer term optimisation. There are already substantial signs that exotic conifers produce a progressive acidification of soils. Ovington and Madgwick

(1957) have shown that Douglas fir (*Pseudotsuga taxifolia*) and Western Hemlock (*Tsuga heterophylla*) produce an exceedingly acid litter. This is particularly unfortunate as both species are extremely high yielding and are frequently planted at low elevation, former agricultural land typified by soils of the Brown Earth categories. Thus site degradation of inherently fertile soils is occurring. Rennie (1962) has reported both chemical and physical deterioration of a soil following afforestation but the rates of change are both slow and somewhat variable. It appears essential that extensive monitoring of forest production must be improved and that pedogenic processes must also be constantly considered alongside timber production figures so that any deterioration in the forest system can be identified and rectified as soon as possible.

CHAPTER 6
ASSESSMENT OF GRASSLAND AND AGRICULTURAL PRODUCTIVITY

6.1 THE ALTERNATIVES TO FOREST VEGETATION

Considerable scope and attention has been given in earlier chapters to the measurement of forest production values. This approach was followed because forest vegetation appears to be the ultimate vegetation type (the climax vegetation) for most parts of this planet. However, a forest cover and agricultural production are not compatible and forest clearance has gone on apace throughout the world. Following deforestation the energy and nutrient patterns at a site are greatly altered. The increase in solar radiation which reaches ground level in a deforested site frequently attains 100 per cent whereas in a closed-forest community, ground level radiation levels may be as low as 8 per cent when expressed as a percentage of above-crown radiation intensity (Baumgartner, 1967). This surge of radiant energy encourages a host of light-demanding species to invade the area, such as the Hawkbit (*Leontodon taraxocoides*) and Ribwort Plantain (*Plantago lanceolata*) while previously shade-tolerant species may show a schizophrenic response and change from small, infrequent components of the vegetation to large, dominant species. Good examples of this can be found in the spread of bracken (*Pteridium aquilinium*), or blackberry (*Rubus fruticosus*) following forest clearance.

The removal of a forest cover inevitably results in an increase of grasses, shrubs and herbaceous plants. The grass family, *Graminae*, is probably the most successful of any single plant family on this planet in that it has successfully invaded every conceivable habitat. Arid or wet, elevated, exposed, sheltered, shaded, saline, alkaline or acid site has fallen to the colonising grass plant. Man has extended its usefulness even further in that he has used the family for extensive plant breeding so that all our modern cereals are highly sophisticated members of the grass family. Modern man has based many of his agricultural systems upon the grass plant. Pastural farming systems, for example, are based upon a continuous grass crop and upon which herbivores (cattle, sheep, goats, horses) are allowed to graze. At varying intervals, between 3 and 10 years, the grass sward will be ploughed into the soil and reseeding will occur. The seed mixture is unlikely to be composed of only one grass variety. Instead a grass–legume mixture will be sown in which three or four grass varieties will be present along with nitrogen fixing legumes such as vetch (*Vicia* spp.) or red and white clover (*Trifolium pratense, T. repens*). The species mixture is chosen for its ability to survive the soil and climatic conditions operating at a site as well as the grazing pressure the farmer expects the field to receive. As a result of the large areas of the world now covered by agricultural grassland and its considerable contribution to food production the study of its primary production has received considerable treatment.

An alternative agricultural land use is that of arable farming in which the soil receives varying degrees of treatment, i.e. ploughing, discing, harrowing, drainage, fertilizing and in which a crop, other than grass, is planted. This crop may be a cereal-grain crop such as wheat, maize, rye, barley, or it may be a root-crop, sugar beet, turnip, swede, or a leaf-crop – usually one of the brassica family. Add to this range of crops those species which have an industrial use as opposed to a food use, such as ground nuts, cotton, tobacco, only then does the true variety of agricultural crops begin to emerge. The monitoring of production rates of all these crops is of the utmost importance for two reasons. Modern agriculture can be considered to be an industry in which the profitability of an enterprise assumes very considerable importance. The farmer, therefore, has to know the cost of producing a crop and yield (production amount), and must also be prepared to investigate the cost of increasing yield through improvement of site conditions or by improved management inputs. The second reason for calculating production levels of crops is that world population is continually increasing and as such makes increasing demands upon world food supply. Organisations such as the FAO are desperately concerned that the additional population will be fed at least with the basic amounts of foodstuffs. Only if the total agricultural production level is known can an estimate be made of the amount of food to be shared between total population. Unfortunately, the food supply is not 'shared' evenly among all peoples. It has been estimated that 2 000 million of the world's population is underfed and that 500 million are verging upon starvation level (Ehrlich and Ehrlich, 1970). By contrast, the developed world consumes by far the greatest amount of foodstuffs and yet fails to contribute her rightful share of production.

This chapter examines the measurement of productivity rate and production amount from natural or semi-natural grasslands and finally examines a brief selection of agricultural systems.

6.2 THE MEASUREMENT OF PRIMARY PRODUCTION OF NON-AGRICULTURAL GRASSLANDS

The methodology to be used in the calculation of production from non-agricultural (i.e. natural or semi-natural) grassland bears a strong resemblance to the techniques described in Chapter 4 for the measurement of natural woodland production. For example, the same formula for Net Primary Production (N.P.P.) can be used:

$$Pn = \Delta B + L + G$$

This equation was explained in section 4.3. Measurement of grassland N.P.P. via this method does produce one very real problem which was not met with in the measurement of woodland N.P.P. Symbol G, plant losses by consumer organisms, that is the eating of grass by herbivores, assumes very considerable proportions in most grassland ecosystems. It must be emphasised that most grassland-dominated systems are maintained by a considerable grazing pressure from either wild or domesticated animals. As such the consumption by the herbivores can represent a major loss from the system particularly if the herbivores are large and are capable of movement over a wide area. In such a situation the harvest method for calculating N.P.P. must be modified to take account of herbage loss via grazing animals.

A similarity between the measurement of woodland and grassland production concerns the shape and size of the field study area. The approach introduced by Newbould (1970) and described in section 4.2 (Fig. 4.1) may be used for grassland production research though the size of the sample area, buffer zone, measurement area and study area may be reduced from that suggested for forest production work. Milner and Hughes (1968) have suggested that an essential prerequisite of any production assessment would be an accurate description of topography and environment of the area with soil type, geology and climatic inputs all being investigated as thoroughly as time and know-how permit.

The measurement of net primary production of grassland is most easily determined by measuring the biomass of the plant community at both the beginning (time t_1) and end (time t_2) of the experiment. Then by subtracting B_1 from B_2 the incremental gain over time period t_1 to t_2 can be easily calculated. The measurement of biomass varies depending upon grassland type. On reseeded pasture which is intensively managed, it is feasible to cut the pasture at the beginning of the experiment (t_1) and to measure the regrowth to time t_2. Natural or semi-natural grassland or old agricultural grassland does not respond well to such treatment. Indeed, grassland which has not been subjected to regular grazing or mowing pressures may be severely damaged by a heavy cut and regrowth would be unlikely to show a representative response. Reference to the measurement of intensively managed agricultural grassland will be made in section 6.3.

The means of calculating Net Primary Production values (N.P.P.) for non-agricultural grasslands depends upon the successful harvesting of the above ground herbage. This is done by means of clipping and collecting a measured area of grassland at a predetermined height. This process is repeated for a number of sites until a representative set of values is achieved. Considerable research has been given to the problems associated with the calculation of the optimum number, size and shape of the measurement area for the key limiting factor in assessment of grassland production is the considerable amount of time necessary for sample collection and analysis (Mueller-Dombois and Ellenberg, 1974). Certainly, the number of quadrats determines the accuracy of the end result. An aid to estimating the number of quadrats is given by Greig-Smith (1964) who recommends the plotting of a selected mean value against number of quadrats sampled. Thus in Fig. 6.1 the mean weight fluctuates only slightly after some 25 sample quadrats have been taken and thus further samples are not required.

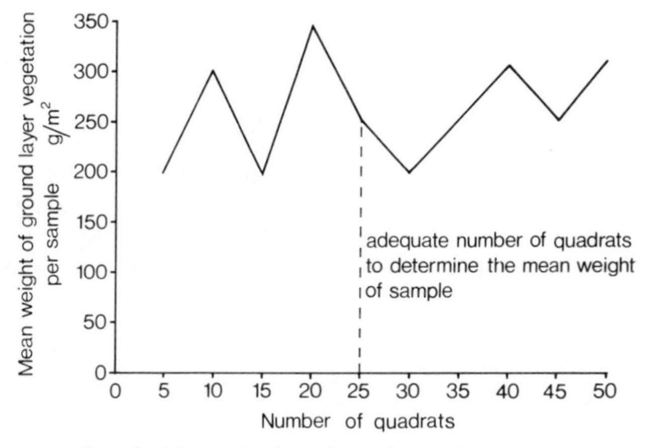

Fig. 6.1 Determination of minimum number of sample quadrats

The cutting of grass within the area of the quadrat is not too difficult a technical problem, but operator error can be a problem. It is necessary to specify and maintain a constant cutting height above ground level and the ease with which this is achieved depends partly upon the species composition of the grassland. Once cut, all the grass must be collected, placed in polythene bags, tightly sealed and returned to the laboratory for immediate analysis. If immediate use of the samples is impossible they should be stored in either a deep-frozen or oven-dried state. Analysis of the cut herbage begins with the weighing of the fresh material and is followed by separation into either the main plant parts, i.e. leaves, stems, flowers, or else separated on a species basis. Account should also be taken of the amount of dead material in the sample, for Weigert and Evans (1964) have shown that grasses produce an appreciable but variable number of shoots of relatively short life. Any production assessment of grassland which ignores the dead material will therefore be an underestimate of the N.P.P. value.

Another measure of particular usefulness in grassland studies is the calculation of the leaf area index (L.A.I.). This measure was defined in Chapter 2 as the ratio of square metres of leaf surface area over one square metre of ground surface. Its calculation for tree leaves is slow but for the fine bladed grasses the calculations become exceedingly laborious and open to much error. Many attempts at automating the calculation of L.A.I. have been made (Orchard, 1961) but the approach of Kemp (1960) has proved highly successful in grassland studies. The method calculates leaf area by the equation:

$$A = kLB$$

where A is area, k is a constant which has to be determined for each species under study, L is leaf length and B is breadth at leaf midpoint.

Determination of grass-root production is yet another major problem in the calculation of N.P.P. and yet root tissue can form more than 50 per cent of the total N.P.P. of some grassland ecosystems (Struik, 1967). Separation of fine roots from the soil matrix is exceedingly difficult as is the apportioning of alive and dead root tissue. Very little work has been done upon root production amounts of non-cultivated species but the work of Weaver (1958) upon natural North American grassland species in which soil monoliths, complete with vegetation and root systems, were washed and analysed, show some of the problems and care necessary in producing a representative result.

Reference to the basic formula $Pn = \Delta B + L + G$ for calculating N.P.P. will show that only variable G (plant losses by herbivores) remains to be considered. The consumption of herbage by large animals (sheep, cows) can be calculated quite easily (see section 6.4a). It is the small wild fauna that shall be considered first. The proportion of herbage eaten by small rodents, rabbits, voles and by the invertebrate fauna is not fully known; again relatively little work on non-cultivated grasslands has been made. Comparison with agricultural grasslands has suggested that cows allowed to graze freely in a field consume about 15 per cent of the N.P.P. (Macfadyen, 1964) and that the remainder is channelled into the wild fauna and decomposer groups. This example suggests that grazing consumption by wild animals and the activities of decomposers are of very considerable significance and require careful monitoring in grassland production studies.

The calculation of grassland N.P.P. values via harvesting of material from a number of quadrats and over an entire growing season is, consequently, fraught with problems. There are some alternative methods which do not require harvesting of material but unfortunately there are other problems associated with their use. As the N.P.P. value represents the carbohydrate material manufactured by the green plant via photosynthesis minus respiration requirements it should be feasible to measure production levels by the uptake of CO_2 which in turn is used in photosynthesis. Billings *et al.* (1966) have shown that it is possible to monitor uptake of CO_2 in the field using a continuous infra-red gas analysis technique. However, this process involves encapsuling a single shoot, or at best a single plant, within a respiration chamber. It is not known with any certainty how conditions within the chamber approximate real-life conditions nor indeed how CO_2 uptake in single plants or in parts of plants equates with an entire grassland quadrat. An alternative to the measurement of CO_2 uptake is the use of radioactive tracer materials, in particular C^{14}. This process involves placing an airtight chamber around a single plant and then introducing a $C^{14}O_2$/air mixture. The plant is allowed to photosynthesise for a given time after which it is harvested and the amount of radioactive material which has been incorporated into the plant tissue is then measured. The radioactivity level is then calibrated against N.P.P. thereby allowing an accurate estimate of net photosynthesis. A similar method can be used for monitoring root productivity though radioactive phosphorus^{32} appears more suitable than C^{14}. Milner and Hughes (1968) give more detailed accounts and references of methods relating to the use of both C^{14} and P^{32} as tracer materials.

A final example of non-destructive sampling can be found in the somewhat simple 'incremental method'. This involves the repeated measurement, over a period of time, of the growth of a plant. New Zealand ecologists have found this method of particular use on the tussock grasslands (Mark, 1965) where increases in tussock diameter, over time, have been used to indicate both vigour of growth and production level.

6.3 THE MEASUREMENT OF THE PRIMARY PRODUCTION OF SHRUBBY PLANTS

Few non-agricultural grasslands are composed of 100 per cent pure grass species. There will inevitably be small flowering plants such as bedstraws (*Galium* spp.) and the yellow tormentil (*Potentilla erecta*). There may also be bracken (*Pteridium aquilinium*) if the grassland was formerly in agricultural use. Northern and/or oceanic locations for grasslands are usually associated with heath-type vegetation characterised by common heather (*Calluna vulgaris*), the purple heather (*Erica cinerea*), cross-leaved heath (*Erica tetralix*) and the bilberry (*Vaccinium myrtillus*). Sampling of these vegetation types can follow the established practice outlined in the preceding section. The heath species mentioned above have the advantage of producing woody tissue with the formation of annual rings. Thus age-determination and annual incremental growth rate and amount can be calculated using methods akin to those described in section 4.3. Harvesting of heath vegetation is comparatively easy; the above ground vegetation can be easily cut away, leaving a ground surface with only a minimal vegetation cover. (The density of ground flora will be dependent upon the luxuriance of the heath cover; old heath, i.e. 15–20 years, often gives a dense tangle of plants which prevents light from penetrating to the floor.) Heath vegetation can support a variety of animal life and thus the grazing pressure can, under some circumstances, be heavy. Heathland in eastern Scotland, for example, can carry a considerable red-grouse population which graze the vegetation. Losses due to herbivores must therefore be assessed with some care and wire cages are often used to eliminate this grazing pressure. Further consideration of cages is given in section 6.4a.

Litter fall in heath communities can be an appreciable amount and must be measured by collecting trays placed beneath the heath canopy. Cormack and Gimingham (1964) have shown that tautly stretched stockinette can trap litter where it falls and this enables litter to be apportioned to individual plants.

The main problems associated with measuring N.P.P. of natural or non-agricultural grasslands is that often they occupy small, inaccessible areas which, in turn, makes field work difficult. Also, natural grasslands are nowadays confined to the least productive parts of the earth's surface (agricultural or improved grassland having replaced it) and thus growth increment, i.e. productivity, can be low particularly in relation to the total N.P.P. Furthermore, variability between sites is considerable and this requires a greater number of samples which in turn increases the amount of field work. Because of the variability of conditions no one approach to grassland with or without shrubby plants can be considered ideal.

6.4 THE MEASUREMENT OF PRIMARY PRODUCTION OF AGRICULTURAL OR IMPROVED GRASSLANDS

The nature of agricultural or improved grasslands can be exceedingly varied depending upon a host of factors, the most important of which are climate, soil, topography and the influence of man. In old agricultural grasslands it may be difficult to recognise the influence of man, for with the passage of time all 'improved' grasslands tend to revert to natural conditions. Indeed, the term grassland is something of a misnomer in that modern agricultural grasslands usually contain more non-grass species than grass species. Non-grass weed species such as daisy (*Bellis perennis*) or chickweed (*Stellaria media*) could well be present while leguminous plants are now invariably contained in grassland mixtures because of their ability to contribute nitrates to the plant component of the system. For the purpose of this book, therefore, the term agricultural grassland will be taken to include sown grasslands (leys) with an abundance of grass and legume species and a marked absence of herbs, shrubs and trees. Another feature of sown grasslands when compared with non-agricultural grasslands is their frequent simplicity in terms of botanic composition; only rarely do they match the floral diversity of natural grasslands.

Grassland vegetation has been used by man as the basis for supporting domesticated animals for many thousands of years. At first the efficiency of the plant-grazing animal system was poor, partly due to the unimproved nature of the grassland, partly to the poorly bred nature of the animal stock and partly also to man's poor understanding of grazing pressures, soil exhaustion and nutritional requirements of his animals. These factors still apply to many of the under-developed areas of the world where farmers are caught in the vicious problems inherent in subsistence farming when the 'one more crop' philosphy tends to outweigh the longer term agricultural-stability approach. In developed countries, however, the efficiency of pastoral farming has reached exceedingly high levels and relatively constant levels of production are attained. Set into a broader, overall picture of plant production, however, agricultural grassland does not rank very highly in terms of production output. There are many reasons for this. First, when compared even to low growing, scrub-woodland, the total plant biomass of a grassland is less than that of the woodland. Forests tend to be multilayered while grasslands tend not to be so. Grasslands also have a very substantial proportion of its biomass locked up, and hence unavailable to grazing animals, in the form of root mass. Troughton (1957) has provided figures which show that *Bromus marginatus* has between 33 per cent and 52 per cent of its total biomass below ground while for *Agropyron smithii* this figure reaches 64 per cent. Grasslands are also very short lived by comparison to woody species. Whereas the latter grow over a period of many years the longevity of grasses is very much shorter. Complete grass swards may last for decades though the individual leaves and roots seldom survive longer than one year and usually less than eight weeks (Spedding, 1971). Because of the absence of lignin the grass shoots and roots decompose quite quickly and this again contrasts to the frequent delay and accumulation of debris in many forest ecosystems.

Although the agricultural grass sward has been described as a relatively simple structured and botanically impoverished vegetation unit, the measurement of grassland productivity is still fraught with difficulties. The general approach to the calculation of productivity can utilise a methodology similar to that described for non-agricultural grassland (section 6.2) although the following modifications may be necessary.

(a) Grazing by large herbivores

The purpose of agricultural grassland is to provide a regular source of food for cattle and sheep and to a lesser extent to horses, goats, pigs and poultry. The grassland species composition will have been chosen to withstand a grazing pressure; indeed, the continued removal of grass leaves by nibbling will stimulate the tillering rate of the grass plant (see section 1.2). In grazed areas, however, one of the commonest techniques has been to cover over an area of grassland by a wire cage so as to exclude grazing animals and thereby a measure of true vegetation productivity can be obtained. The exclusion of grazing animals places a totally different set of environmental conditions upon the grass community and N.P.P. values are unlikely, therefore, to provide a true estimate of production (Welch and Rawes, 1964). The effect of exclusion can be minimised provided that the duration of exclosure is kept short relative to the total life cycle of the main species. As grass species are short lived (see above) then exclosure experiments should, in theory, be confined to two- or three-week durations. Local field conditions will determine the exact duration of the experiment as enough time must be allowed for sufficient growth to take place and be measured so that a change in herbage biomass can be recorded. The effect of cages on herbage production can vary though exclosures always permit an increase in N.P.P. values when compared to a 'control' site. Thus the protective role of a cage seems to exceed the stimulus produced from a grazing pressure. Dry matter production usually increases by between 10 and 15 per cent (Cowlishaw, 1951) though in high rainfall areas the increase is frequently lower. The difference between production values outside and inside the cage will represent the net primary aerial production (N.P.A.P.). The effect of exclosures on root systems is not known. Once the N.P.A.P. value is known it is then relatively simple to show that:

$$\text{Herbivorous intake} = \text{Amount of herbage produced within the cage} - \text{Amount of herbage produced outside the cage}$$

This formula does overestimate somewhat the herbivorous intake due to the increased relative growth rate within the cage. The above formula is applicable only to the most simple situations and Milner and Hughes (1968) give modifications of the formula to suit more complex conditions.

(b) Application of inorganic fertilizers

One of the major differences between agricultural and non-agricultural grasslands is the application of chemical fertilizers to the former in an attempt to maintain or improve the capacity of the sward to support grazing herbivores. Any ploughing of old grassland and subsequent reseeding with modern grass mixtures inevitably involves also an application of basic slag (to control any deterioration in soil acidity) and/or application of a nitrogenous fertilizer. The impact of fertilizers upon the sward is to produce an energy surge through the system and an improved growth of the grass crop usually occurs. Not all swards respond equally to fertilizer treatment; there is certainly an optimum limit of fertilizer application beyond which improvement in plant yield shows a limited response. Work by Reid (1966) showed that for swards comprising mainly

Lolium perenne the relation between dry-matter yield per annum (Y_{dm} in 100 lb/acre) and nitrogen application (= X in lb/acre) followed the relationship:

$$Y_{dm} = 123{\cdot}18 - 69{\cdot}63e^{-0{\cdot}00130X^{1.32}}$$

where e is the base of natural logarithms. In practice this meant that when up to 150 lb/acre N was added a response of 26 lb dry matter/lb N was achieved. The response gradually declined so that application rates between 200 and 250 lb/acre N gave a response of 13 lb dry matter/lb N and 500 lb/acre N gave 1 lb dry matter/lb N.

All calculations of agricultural grassland production must take into account the impact of fertilizer applications. Response curves should be drawn for increase in yield plotted against fertilizer amount but taking into account also the decline in fertilizer effect with the passage of time.

(c) Irrigation

In many parts of the world grassland production is limited mainly by non-availability of water. Much Australian grassland is thus limited. The effect of irrigation on grass yield is similar to that of nitrogenous applications reviewed in (b) above. In Britain, where grassland irrigation is rare, a soil moisture deficit of 37 mm results in a 5 per cent decline in grass yield while a 50 mm water deficit produces a 10 per cent decline (Tayler, 1965). Factor interaction, however, plays a significant role upon grass yield, for Spedding (1971) has discussed the inter-active effects of light, nitrogen, water, carbon and time upon grass yield.

The factor of time is, perhaps, the most elusive of all. Any study of plant production inevitably requires the passage of time over which the growth rate and amount can be measured. The presence of measuring equipment and the methods adopted for measuring can change the environmental inputs to such an extent that the results become unrepresentative. Agricultural grasslands which at first sight appear simplistic are really infinitely complex because of numerous levels of interaction which are possible between man and natural environmental variables.

6.5 THE USE OF MODELS IN GRASSLAND PRODUCTION STUDIES

Reference has already been made in section 5.5 to the desirability of modelling plant growth with particular reference to the different management techniques which can be applied to forest growth. Because of the complexity of forest ecosystems it is exceedingly unlikely that a workable forest model can be produced within the next 10 years or so. A grassland model is much more of a reality because the relative simplicity of a grassland ecosystem allows the main parameters to be identified and measured and the volume of agricultural research greatly exceeds forest research mainly because the need for food is more apparent than is the need for timber. An example of a dynamic model (i.e. one capable of incorporating changes as a result of feedback as the model is operated) has been described by N. R. Brockington and quoted by Spedding (1971). Brockington's model considers the problem of pasture contamination from the excreta of grazing cattle. As with De Wit's ELCROS program, Brockington's model simulates reality by advancing through time in small discrete steps and in which the relationships between the inputs are expressed as differential equations. Figure 6.2 shows the problem of pasture contamination in flow-diagram form. The rectangular box designated 'Area contaminated' is the only level, or quantity which can be physically measured. Feeding into the box is a solid line which symbolises a physical flow of

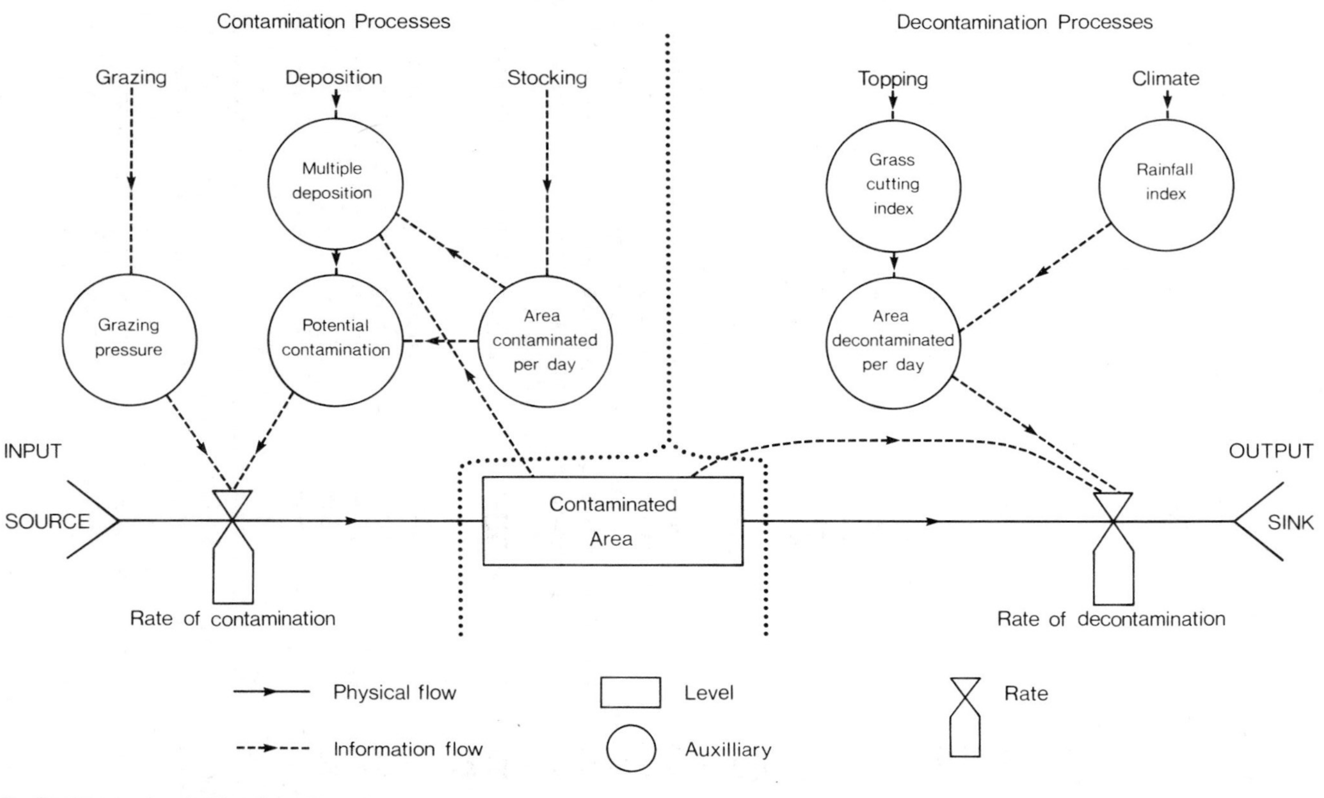

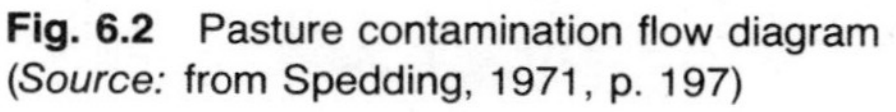

Fig. 6.2 Pasture contamination flow diagram
(*Source:* from Spedding, 1971, p. 197)

material. On the left of the 'Area contaminated' comes a flow from a 'source' area which is of unknown and changeable dimensions. To the right of the 'Area contaminated' may be found another flow to a sink area. Again the dimensions of the sink, its capacity and speed of intake, are unknown. The rate of flow from source through to sink area can be represented by symbols equating to valves. The dashed lines represent flow lines of information to valves, originating from the 'level' box and both to and from the final components of the model, the auxiliaries, depicted by circles in the diagram. The auxiliaries provide, as the name suggests, extra information, the more of which are available help improve the reality of the model. In Fig. 6.2 it is possible to investigate the processes of contamination and decontamination as distinct functions, the former having as a prime variable the stocking rate, i.e. the number of cows per unit of pasture area, and the decontamination process being primarily under the control of rainfall frequency and amount although the 'topping' rate – i.e. the amount and frequency with which the pasture is cut between grazing periods – will also influence decontamination rate.

Next, the model has to be transferred to a mathematical language suitable for computer use. The example quoted here has been operated via a simulation language called DYNAMO though the more frequently used FORTRAN-based BASIC language could also be used. The model rarely allows absolute results to be produced. Instead, simulations can be made using a variety of different auxiliaries. In the pasture contamination model, for example, some substantial gaps in knowledge are revealed which require further research. Thus, a 'grey' area exists over the influence of animal stocking rate on the proportion of contamination per day. This model is an example of a relatively straightforward problem but the scope for further development even of this case study is immense. Comparable models could be built of factors controlling the growth processes of a selected crop or production patterns of a plant or animal unit and ultimately the models could be interlinked to give a new understanding of the processes which control 'agricultural ecosystems'. Such a goal is still a long way off and instead we must be satisfied with the methods described in earlier sections of this book for the measurement of productivity, production and biomass values of plants.

CHAPTER 7
CONCLUSION

7.1 THE BROADER IMPLICATIONS

Most of this book has been concerned with examining the methods which can be used to monitor vegetation productivity. Enough has been written to show that the measurement of plant growth is a slow, tedious and often inaccurate process and only a very few thorough assessments have so far been made, notably Whittaker and Woodwell (1968) and Duvigneaud (1971). But the measurement of vegetation productivity on its own represents a very small part of the complete attempt by man to achieve an ensured supply of vegetable materials in the final years of the twentieth century. Burnet (1966) has stated that man is now faced with three major concerns: 'to reduce war to a minimum; to stabilize human population; and to prevent the progressive destruction of the earth's irreplaceable resources'. Vegetation is by no means irreplaceable; there are, as yet, no fears of vegetation disappearing from the surface of the earth and yet some types of vegetation, the truly natural vegetation units, the so-called climax vegetation types have, almost without exception, already disappeared. The vast deciduous forests of north-west Europe and of the eastern seaboard of North America, the mixed forests surrounding the Mediterranean Sea, the Redwood forest of the northern Californian coast, the African Rain Forest have either completely disappeared or have been reduced to extremely small remnants. The list of examples is endless for wherever man has established himself in significant numbers then so follows 'progressive destruction of the earth's irreplaceable resources'.

Natural vegetation is just one of the biotic resources located within the biosphere. Along with the abiotic resources of soil, air and water, it establishes the foundation for the well-being of the entire animal kingdom for it is the photosynthetic capacity of green plants which in turn allows man to reach his present world population total of about 4 000 million persons. Each man, woman and child is dependent upon the ability of plants to generate chemical energy from an input of radiant energy plus inorganic substances. We can assume that the total radiant energy received by this planet is both fixed and finite in amount. As man's total population increases and as our standard of living improves so our demand for plant-derived foodstuffs increases and, therefore, by implication man is diverting more and more vegetation production, and hence energy, towards the support of one species, *Homo sapiens*. The fundamental question to be posed is: Can man attain a dynamic equilibrium with the biosphere in order that an imbalance does not occur? The nature of the potential imbalance can be many and varied; some which have attracted considerable concern in recent years are the contamination by pollutants of air, land and water. Any decline in the quality of the environmental components would ultimately produce an impaired biosphere in which vegetation productivity would be likely to decline. A less obvious effect of man upon the biosphere but one with equally great impact is his ability, indeed his active organisation, to cause simplification of natural ecosystems. The extreme complexity of measuring the productivity of a natural forest was shown in Chapter 4. A multitude of tree species with up to four or five distinct layers of vegetation, a wide range of age groups, and a staggering of life histories all combine to give an intricate mosaic of vegetation communities. The inter-relation of communities in natural vegetation units provides stability unknown in man-managed systems. Elton (1971) has discussed some of the reasons for the differences in stability levels. He suggests that if a biologic community be represented by a mathematical model of 'n' species, when 'n' equals 2 there will be direct competition between the two species and the stronger will eventually dominate. When 'n' equals a much larger number of species, for example 20, then complex interactions occur between the species with the formation of a food web and niche patterns becoming evident. Under these circumstances no single species is likely to emerge as the overall dominant species and, similarly, no species is made extinct. In reality, biologic communities are infinitely complex with number of species being unlimited. While the content of this book has been concerned only with vegetation production it must be stressed that animal populations can also be assessed for production values. The superimposition of plant and animal communities provides yet further stability as feeding chains which involve both organic groups bond the community even more closely together.

Natural organic systems have proved exceedingly rich resource areas within which man has evolved. As modern man's capabilities as a farmer, forester, industrialist or economist have increased then gradually more and more species were utilised from the resource base. Ultimately, man created his own resource base – that of the modern farm. The most contemporary agricultural unit attempts to optimise the flow of energy from the sun through the soil and through plants and animals into an organic form which is most suitable for man. Optimum production can be attained by man only when the system is grossly simplified, for example, a field of wheat or a herd of Hereford cattle. In creating a simplified system in which the number of species is extremely low we return to Elton's example of an inherently unstable model which is difficult to maintain.

The simplified agricultural system is perpetrated by man only with considerable input of energy. Of great significance is the use of synthetic chemicals with which to control insect and weed pests. Mellanby (1967) has provided a brief review of the evolution of herbicides and insecticides. Substances such as M.C.P.A. and 2,4-D are the most widely used in Britain though there are about forty other chemical compounds in regular use. M.C.P.A. is an 'auxin-type growth regulator' (see section 1.3) in that when applied to herbaceous weeds it causes excessive growth and eventual rupture of the cell walls leading to a disintegration of the plant. Most of the modern chemical additives are described as 'non-toxic' but this term is only relative. The widespread use of D.D.T. from 1939 onwards reduced beyond all imagination the possibility of disease transmitted by lice. D.D.T. was hailed as the wonder substance and its use was extended to every point of the globe. Gradually, however, its effectiveness appeared to be impaired; flies which once died from a single application of D.D.T. now required several applications. Insect resistance had been achieved possibly by genetic modification. Then it was discovered that D.D.T. could be accumulated in the body, particularly in fatty tissue. Penguins in the Antarctic were found to be contaminated with D.D.T. even though they had never been directly sprayed with the substance as was the case also for many species of fish dredged from deep waters. The Canadian Government was the first body to ban the use of D.D.T. (in 1970) because of its non-degradable qualities. This step was taken after considerable laboratory testing at various centres throughout North America and also in Sweden. From the latter country came the fearsome recommendation that newly-born babies should not be fed on natural mother's milk because it may contain D.D.T. concentrations high enough to produce digestive complications (Crowe, 1972).

The natural ecosystem dominated by natural vegetation units has been replaced by the so-called modified ecosystem (Billings, 1972) and yet as has been shown in the previous paragraphs the modified ecosystems are extremely unstable. In the densely peopled parts of the world with a long history of settlement, urbanisation and industrialisation there has already occurred considerable ecosystem damage. Soil erosion in the Steppe Lands of Russia, the lowering of the water table as in the London Basin, the destruction of the citrus fruit industry in parts of California from photochemical smog are all examples of ecosystem damage. One of the greatest tests of modern man's technologic and scientific skills still remains to be achieved. That is the resynthesis of ecosystems in which natural homeostatic mechanisms are recreated. Some ecosystem resynthesis has admittedly been achieved already; for example, the highly successful Tennessee Valley Scheme (Clapp, 1955) or the return of industrially scarred land to agriculture or recreational land use as in much of the counties of Lancashire and County Durham in England (Arvill, 1967). Bates (1969) has emphasised the prudence in retaining as many of the natural ecosystems as possible at least until man's skill as an ecosystem manager has improved to the extent that ecosystem repair and resynthesis becomes the norm and not the exception. It is only the natural ecosystems which have within them the blue-prints for the very evolution of the future of man on this planet. The measurement of vegetation productivity represents just a beginning, just one stage, in the collection of objective data and understanding of vegetation units. There is little doubt that while ecologists and agriculturalists understand part, and realise that there is still much more to be learned, of plants and their growth patterns, immense pressure is placed upon them to raise productivity levels even further. Sufficient has been said in the earlier chapters of this book to show that plant growth is under the control of a number of processes many of which are beyond the control of man. Several other facts have also to be recognised: that the most easily utilised and at the same time productive regions of this planet have already been converted from natural vegetation ecosystems to modified agricultural systems; that the really high natural vegetation productivity regions (those within the Tropics) are by no means easy to convert to stable, modified systems and that thirdly in some parts of the world, notably the heavily industrialised northern hemisphere, European countries, population densities are so great that it is impossible to maintain them with home-produced food supplies.

Pressurised by governments, relief organisations and international bodies such as World Health Organisation, the exploitation of vegetation systems has gone on apace and with little regard for sensible, ecologically-based development. Slesser (1975) has provided data to show that for many agricultural enterprises (extreme examples of modified ecosystems), there has been a progressive decline in plant productivity relative to the input of energy from anthropogenic resources. Heichel (1973) in a study of energy effectiveness has shown that in Minnesota many oat farms and in Missouri the soy bean farms produced the same quantity of digestible energy in 1973 as did maize farms in those states in 1915 but that the modern day farm has a higher energy input. Heichel's results should be treated with caution as his values are based upon financial conversion ratios but the trend of his results appear correct for it has been repeated by other research projects in different areas (Pimental, 1973). It has been shown in Chapter 3 that agricultural productivity can rarely match natural vegetation productivity in a similar environment and it was suggested that this was due mainly to the relative shortness of the agricultural growing season when compared with a natural ecosystem. This is only partly true. The agricultural system can be considered to be considerably bolstered by additional energy inputs – fertilizers, insecticides, manipulation of the agro-environment by mechanisation and its consequent energy demands. The law of diminishing returns soon sets in as was shown in section 6.4b in respect of fertilizer application and crop yield. Data quoted by Slesser (1975) has shown that in the mid-1950s the substitution of a single hour of a farm-labourer's time on an American farm involved investment in mechanisation which in turn consumed a little over 1 gallon of petrol. By 1970 a further 1 hour saving of man power would demand no less than 4·3 gallons of petrol. Leach (1975) has shown that in Britain for every one job lost on a farm through improved efficiency there comes an increase of about half a man's work in the agri-technology business and a whole man's job in food-processing and delivery service. One is forced to question very seriously whether the trend towards increasing agricultural productivity is indeed a true increase in the efficient use of rapidly diminishing resources.

The twentieth century has been a time in man's development that has been characterised by a revolution of rising expectations (Jaguaribe, 1966). Personal wealth, job expectations, quality and standard of life have all witnessed a tremendous advance particularly in the years from 1960 onwards. Even the impact of a major world economic recession during the mid 1970s was cushioned by the elaborate infrastructure which man has built up in his society, e.g. state unemployment payments, compensation for enforced lay offs, redundancy hand-outs for closure of old, inefficient industries. Geographers have for long acquiesced to the theory that the resources of this planet have been readily available for the use of man. Some attempt was made to show that some resources with a particularly long renewal rate should be termed 'non-renewable', e.g. coal, oil, iron-ore, while others with a rapid reproduction rate were termed 'renewable', e.g. forests, fish stocks. Some resources which were available in exceedingly large supply were also treated with the same attitude as the renewable resources, thus in Britain, fresh clean water was so abundantly available that little concern was given to its long-term supply prospects. But the irreversible and irrepressible needs of economic growth associated with an increasing population with spiralling expectations changed this situation in the space of some 20 years from 1950 to 1970. The cost of purifying polluted water sources and the search for new water supplies has shown that Britain is not as well endowed with a water resource as was once thought.

Vegetation has long been considered a renewable resource. Since the very beginnings of man on this planet this has been so, and indeed the concept was perfectly correct for some 90 per cent of man's history. If one accepts the thesis of Lovejoy and Homan (1965) that the 'resource base' is the total sum of all components of the environment on this earth and that the 'resources' are those parts of the resource base that man can utilise under prevailing technologic, scientific, economic and political conditions, then it follows that as man's ability to utilise the resource base extends then more and more resources become available to him. Mention has been made frequently through this book of man's increasing technical and managerial skills and this has been the factor behind the phenomenal leap forward in the status of mankind on this planet. Fundamental to the problem of resource use is the factor of numbers; the number of men, women and children on this planet. Our population, according to Ehrlich and Ehrlich (1970), can now double its size in 35 years. With such a rate of increase we shall have to achieve a doubling of vegetation production over a similar time span if we are to retain current standards of diet. Such a rate of increase, an average of 10 per cent more production of all plant-derived items every 3½ years, is totally impossible on a world scale. Some countries may achieve it for a short time and for selected products, e.g. France, New Zealand and possibly Brazil, but highly developed and densely populated nations such as Britain have little hope of attaining anywhere near this rate of increase. For forest production, Britain hopes to achieve an approximate 22 per cent increase in output over the next 35 years, that is, 78 per cent below the required output!

What prospects, then, are in store for vegetation patterns on this planet when such massive demands are to be placed upon them? It would be easy to say quite simply that exploitation of ecosystems will occur, that a movement away from ecological optimisation of production to a maximisation of production will happen. Certainly, with our improved agro-technology and agro-sciences some move in this direction will be inevitable but will it be sufficiently great to allow us to achieve a doubling of production in 35 years? It is exceedingly unlikely.

There is, of course, one other possibility and that is that we shall not require a 100 per cent increase in 35 years. Disease, or warfare, might decimate the population; such events are well recognised and unfortunately all too repeatable phenomena in man's history. More hopefully man may reduce his rate of increase as the problems of over-population become felt. Most other plant and animal societies obey this general rule. As population size becomes larger then so further increase is made more difficult by 'environmental resistance'. Man is still a relatively young species on this planet and, possibly, has yet to attain a stabilised population size. Hopefully, also, there will be a change in man's ethical stature in which overcrowding and the associated living squalor will give way to a lower population density which can be supported at a very much higher standard and quality of life by the prevailing technologic and economic conditions. Fundamental changes in the attitudes of mankind will be called for. Hardin (1968) has produced an excellent analogy of man's ethical concepts with that of the mentality of the 'commons'. His argument, much abbreviated, is that our resource base can be likened to that of a plot of common land, that is communal land or land with no formal ownership. In the past, common land was open to all men for whatever purpose they wished – some grazed sheep, cattle or goats, some cut timber or peat while others eventually built their houses on common land. All was well so long as the technological skills did not permit over-exploitation of the common or that the total population using the common did not exceed a critical size. Such an idealised situation probably never existed. Gradually, the individuals who used the common land for grazing of cattle would add to their herd. Isolated additions would have little overall effect on the ecological balance but when each user added a cow then over-exploitation would occur leading to site degradation.

The 'commons attitude' is still all too prevalent among all levels of society today. Man, as an individual, often feels that he or she cannot begin a movement away from the commons attitude unless everyone else does likewise for the fear is that personal materialistic losses will occur if an expansionist economy is replaced by one dominated by permanency. For example, how can one person's decision not to own a car help reduce air pollution levels? A similar approach lies with the use of vegetation. One more tonne of wheat per hectare or one more cubic metre of timber increment from a forest area represents the same 'commons attitude'.

How we begin to move away from such a commons attitude is difficult to tell. Any change in society which allows us to do so is likely to be a slow, erratic process and until it is fully achieved man will continue to demand improved production from plants. It is imperative that we know how much production we can obtain from the different vegetation regions of this planet so that exploitation does not occur. Accepting that natural vegetation communities are the optimum producers automatically sets an upper production limit. By comparing agricultural production with the natural plant production for a similar geographical position we can begin to judge the efficiency of man's use of the biosphere. No matter which economic, political, social, religious or ethical framework may prevail such a comparison is necessary if man is to survive on this planet. The measurement of vegetation productivity is, consequently, of fundamental importance to our continued well being.

GLOSSARY

Angiosperm The most 'modern' of the plant groups. First emerged as distinct group of species some 130×10^6 years ago. Now has the greatest variety of species of all plant forms (300 000 different species). Complex flowers and elaborate protection for seeds. Deciduous and evergreen types. 'Hardwood' trees.

Assimilated matter Material generated by green plants via photosynthesis. Primary production.

Biomass Strictly defined as the total organic matter at a site. Because of measurement difficulties, decomposer group and/or root system often omitted from survey.

Bryophyta The phyllum containing all mosses and liverworts.

Chlamydomonas viridis A primitive uni-cellular flagellate with both plant and animal characteristics, i.e. chloroplasts and movement.

Coniferales Cone-bearing trees. Formerly extensive but now confined to about 550 different species. Resinous, softwood timber. Mostly evergreen trees.

Consumer units A term used by agriculturalists when forecasting future demand for food. A consumer unit represents a predictable demand for amount and type of foodstuffs. The demand level may vary with the social status of the consumer unit.

Encystment A form of protection undertaken by lowly organisms in unfavourable conditions, e.g. heat, drought, salinity. The organism can exist for long periods of time in the state of encystment and at the return of favourable conditions can resume normal life functions.

Exotic tree species Introduced species, i.e. non-native. Usually applied to the man-planted conifer trees now used extensively in commercial afforestation.

Forest mensuration The science of tree growth measurements as related to timber production assessments.

Growing season The period of time from the last killing frost of spring to the first killing frost of autumn. The term is usually applied to the period of the year when visible growth occurs.

Growth potential curve A term used by foresters to predict the expected growth rate of a species over a period of time.

Gymnosperms Seed-bearing trees, with seeds carried in cones. Formerly very extensive (Mesozoic era) but now mainly confined to the Coniferales.

Heartwood The central core of a tree trunk. Composed of xylem tissues. Small, closely packed cells, often stained yellow-brown due to deposition of tannin.

Mycorrhizal infections Fungal growth on either the outside (ecto-) or inside (endo-) of roots. It is thought that the fungal hyphae provide vital nutrients for the plants. Very common on plants which inhabit acidic heaths.

Production A measure of the amount of accumulated organic matter over a given time period, usually expressed as g/m^2. The most common measure of plant growth, particularly for managed crops.

Productivity A measure of energy accumulation over a given time period (the rate). Usually expressed as $joules/m^2$. Very difficult to assess the productivity of complete ecosystems or even single communities within an ecosystem. Most work confined to individual species.

Pteridophyta Spore-bearing plants typified by fossilised tree-ferns, horsetails and club mosses.

Sapwood The outer, softer wood of the tree trunk. Composed of phloem tissue.

Thallophyta Collective name for the group of organisms including algae, fungi and bacteria.

Quality Class A qualitative classification system for assessing tree performance. The quality class is specific to individual tree species.

Yield class The maximum mean annual volume increment, irrespective of age, which a tree is capable of attaining during its lifetime. A quantitative measure of tree growth. The yield class can be universally applied to different tree species.

REFERENCES

Arvill, R. (1967) *Man and Environment,* Penguin Books, England.

Assmann, E. (1970) *The Principles of Forest Yield,* Pergamon Press, Oxford.

Avery, T. E. (1967) *Forest Measurements,* McGraw-Hill Book Co., New York.

Baker, H. G. (1970) *Plants and Civilization,* Macmillan, London.

Bates, M. (1969) The Human Ecosystem, pp. 21–30 in Committee on Resources and Man (eds), *Resources and Man,* W. H. Freeman & Co., San Francisco.

Baumgartner, A. (1967) 'The balance of radiation in the forest and its biological function', pp. 743–54, in Tromp, S. W. and Weihe, W. H. (eds), *Proceedings of International Biometeorological Congress,* Pergamon Press, Oxford.

Bazilivich, N. I., Rodin, L. Y. and **Rozov, N. N.** (1971) 'Geographical aspects of productivity', *Soviet Geography,* **12,** 293–317.

Best, R. H. (1968) 'The extent of urban growth and agricultural displacement in post-war Britain', *Urban Studies,* **5,** 1–23.

Billings, W. D. (1972) *Plants, Man and the Ecosystem,* Macmillan, London.

Billings, W. D., Clebsch, E. E. C. and **Mooney, H. A.** (1966) 'Photosynthesis and respiration rates of Rocky Mountain alpine plants under field conditions', *American Midland Naturalist,* **75,** 34–43.

Blackwell, M. J. (1966) 'Radiation meteorology in relation to field work', pp. 17–39, in Bainbridge, R. (ed.), *Light as an Ecological Factor,* Blackwell Scientific Publications, Oxford.

Bleasdale, J. K. A. (1952) 'Atmospheric pollution and plant growth', *Nature,* **169,** 376–7.

Bormann, F. Herbert and **Likens, Gene E.** (1970) 'The nutrient cycles of an ecosystem', *Scientific American,* **223,** 92–101.

Botkin, D. B., Janak, J. F. and **Wallis, J. R.** (1972) 'Some ecological consequences of a computer model of forest growth', *J. Ecol.,* **60,** 849–72.

Brady, N. C. (1974) *The Nature and Properties of Soils,* Macmillan, London.

Bunce, R. G. H. (1968) 'Biomass and production of trees in a mixed deciduous woodland', *J. Ecol.,* **56,** 759–76.

Burnet, M. (1966) *Ecology and the Appreciation of Life,* The Boyer Lectures, Australian Broadcasting Commission, Ambassador Press, Sydney.

Clapp, G. R. (1955) *The T.V.A. An Approach to the Development of a Region.* University of Chicago Press, Chicago.

Clarke, G. L. (1957) *Elements of Ecology,* Chapman & Hall Ltd, London.

Cohen, J. B. and **Ruston, A. G.** (1911) *Smoke, a Study of Town Air,* Edward Arnold & Co., London.

Commoner, B. (1971) 'Evaluating the biosphere', pp. 50–60, in Detwyler, T. R. (ed.), *Man's Impact on Environment,* McGraw-Hill Book Co., New York.

Cormack, E. and **Gimingham, C. H.** (1964) 'Litter production by *Calluna vulgaris* (L) Hull', *J. Ecol.,* **52,** 285–97.

Cowlishaw, S. J. (1951) 'The effects of sampling cages on the yields of herbage', *J. Brit. Grassl. Soc.,* **6,** 179.

Crowe, B. L. (1972) 'The tragedy of the commons revisited', in Smith, R. L. (ed.), *The Ecology of Man,* Harper & Row, New York.

Cruickshank, J. (1972) *Soil Geography,* David & Charles, Newton Abbot.

Darby, H. C. (1977) *Domesday England,* Cambridge University Press, England.

Darwin, C. (1897) *The Power of Movement in Plants,* D. Appleton & Co., Inc., New York.
Daubenmire, R. F. (1974) *Plants and Environment,* J. Wiley & Sons, London.
Day, W. R. and **Pearce, T. B.** (1936) 'The influence of certain accessory factors on frost injury to forest trees', *Forestry,* **10,** 124–9.
De Wit, C. T., Brouwer, R. and **Penning de Vries, F. W. T.** (1971) 'A dynamic model of plant and crop growth', in Wareing, P. F. and Cooper, J. P. (eds), *Potential Crop Production,* pp. 116–42, Heinemann, London.
Dickinson, G., Mitchel, J. and **Tivy, J.** (1971) 'The application of phytosociological techniques', *S.G.M.,* **87,** 83–102.
Dixon, J. R. (1971) 'An attempt at site assessment for Douglas Fir in Perthshire', *Scottish Forestry,* **25,** 26–33.
Duckham, A. N. and **Masfield, G. B.** (1970) *Farming Systems of the World,* Chatto and Windus, London.
Duvigneaud, P. (ed.), (1971) *Productivity of Forest Ecosystems,* UNESCO, Paris.
Edlin, H. L. (1956) *Trees, Woods and Man,* Collins, London & Glasgow.
Edwards, A. M. and **Wibberley, G. P.** (1971) 'An agricultural land budget for Britain 1965–2000', *Studies in Rural Land Use,* **10,** Wye College, Kent.
Ehrlich, P. R. and **Ehrlich, A. H.** (1970) *Population, Resources, Environment. Issues in Human Ecology,* Freeman Books, San Francisco.
Elton, C. (1971) *Animal Ecology,* Methuen, London.
Eyre, S. R. (1968) *Vegetation and Soils. A World Picture,* Edward Arnold & Co., London.
Fitzpatrick, E. A. (1971) *Pedology: A Systematic Approach to Soil Science,* Oliver and Boyd, Edinburgh.
Flannery, K. V. (1969) 'Origins and ecological effects of early domestication in Iran and the near East', in Ucko, P. J. and Dimbleby, G. W. (eds), *The Domestication and Exploitation of Plants and Animals,* pp. 73–100, Duckworth, London.
Food & Agriculture Organisation (annually), *Yearbook of Forest Products,* F.A.O., Rome.
Food & Agriculture Organisation (annually), *Production Yearbook,* F.A.O., Rome.
Ford, E. D. (1971) 'The potential production of forest crops', in Wareing, P. F. and Cooper, J. P., *Potential Crop Production,* pp. 172–86, Heinemann, London.
Ford, E. D. and **Fraser, A. I.** (1968) 'The concept of actual and potential production as an aid to forest management', *Forestry,* **41,** 175–81.
Forestry Commission (1946) 'Spring frosts', *Bulletin No. 18,* H.M.S.O.
Forestry Commission (1954) 'Tree root development on upland heaths', *Bulletin No. 21,* H.M.S.O., London.
Forestry Commission (1963) 'Forest drainage', *Research Branch Paper,* No. 26, H.M.S.O., London.
Forestry Commission (1966) 'Forest management tables', *Handbook* No. 16, H.M.S.O., London.
Forestry Commission (1971) 'Forest management tables (metric)', *Booklet* No. 34, H.M.S.O., London.
Galston, A. W. (1964) *The Life of the Green Plant,* Prentice-Hall, Inc., Englewood Cliffs, New Jersey.
Geiger, R. (1957) *The Climate Near the Ground,* Harvard University Press, Cambridge, Mass.
Glinka, K. D. (1931) *Treatise on Soil Science,* Israel Programme for Scientific Translations (1963), Jerusalem.
Greig-Smith, P. (1964) *Quantitative Plant Ecology,* Butterworths, London.
Gunkel, J. E. (1957) 'The effects of ionizing radiations on plants: morphological effects', *Quart. Rev. Biol.,* **32,** 46–56.
Hardin, G. (1968) 'The tragedy of the commons', *Science,* **162,** 1243–8.
Hatcher, E. S. J. (1959) 'Auxin relations to the woody shoot', *Ann. Bot. (Lond.),* **23,** 409–23.
Heichel, G. (1973) 'Report of the Connecticut Agricultural Experimental Station', Box 1106, New Haven, Connecticut.
Hogg, W. H. (1959) 'Shelter in relation to horticulture', in Taylor, J. A. (ed.), *Shelter Problems in Relation to Crop and Animal Husbandry,* pp. 6–10, U.C.W., Aberystwyth.
Hole, F. and **Flannery, K. V.** (1967) 'The pre-history of S.E. Iran: a preliminary report', *Proc. Prehist. Soc.,* **33,** 147–206.
Huxley, J. (1970) 'The world population problem', in Love, G. A. and Love, R. M. (eds), *Ecological Crisis,* pp. 61–82, Harcourt Brace Jovanovich, Inc., New York.
Jaguaribe, H. (1966) 'World order rationality and socio-economic development', *Daedalus,* **95,** 607–26.
Janzen, D. H. (1975) *Ecology of Plants in the Tropics,* Edward Arnold & Co., London.

Jones, G. E. (1972) *The Effects of Selected Environmental Hazards on the Growth of Picea sitchensis in three Forests in Wales,* Ph.D. Thesis, U.C.W.

Kemp, C. D. (1960) 'Methods of estimating the leaf area of grasses from linear measurements', *Ann. Bot. (Lond.),* **24**(96), 491–503.

Kershaw, K. A. (1964) *Quantitative and Dynamic Ecology,* Edward Arnold & Co., London.

Kimura, M. (1963) 'Dynamics of vegetation in relation to soil development in northern Yatsugadake Mountains', *Jap. J. Bot.,* **18,** 255–87.

Kira, T. and **Ogawa, H.** (1971) 'Assessment of primary production in tropical and equatorial forests', in Duvigneaud, P. (ed.), *Productivity of Forest Ecosystems,* pp. 309–21, UNESCO, Paris.

Kira, T. and **Shidei, I.** (1967) 'Primary production and turnover of organic matter in different forest ecosystems of the Western Pacific', *Japan J. Ecol.,* **17,** 70–87.

Klinge, H. (1975) 'Biomass and structure in a central Amazon rainforest', in Golley, F. B. (ed.), *Tropical Ecological Systems,* pp. 115–22, Springer-Verlag, Berlin.

Kittredge, J. (1944) 'Estimation of the amount of foliage of trees and stands', *J. For.,* **42,** 905–12.

Kurosawa, E. (1926) 'Experimental studies on the secretion of *Fusarium heterosporum* on rice plants', *Trans. Nat. Hist. Soc. Formosa.,* **16,** 213–27.

Leach, G. (1975) 'The energy costs of food production', in Steele, F. (ed.), *Man Food Equation,* pp. 139–63, Academic Press, London.

Leopold, A. C. (1964) *Plant Growth and Development,* McGraw-Hill, New York.

Lieth, H. (1968) 'The determination of plant dry matter production with special emphasis on the underground parts', *Copenhagen Symposium Report,* pp, 179–86, UNESCO, Paris.

Lieth, H. (1973) 'Primary production: terrestrial ecosystems', *Human Ecology,* **1,** 303–32.

Lovejoy, W. F. and **Homan, P. T.** (1965) *Methods of Estimating Resources of Crude Oil, Natural Gas and Natural Gas Liquids,* Resources for the future, Johns Hopkins Press, Baltimore.

Macfadyen, A. (1964) 'Energy flow in ecosystems and its exploitation by grazing', in Crisp, D. J. (ed.), *Grazing in Terrestrial and Marine Environments,* pp. 3–20, Blackwell Scientific Publications, Oxford.

Madgwick, H. A. I. (1970) 'Biomass and productivity models of forest canopies', in Reichle, D. E. (ed.), *Analysis of Temperate Forest Ecosystems,* pp, 47–54, Chapman and Hall, London.

Mark, A. F. (1965) 'The environment and growth rate of narrow leaved snow tussock in Otago', *N.Z. Journ. Bot.,* **3**(2).

Mellanby, K. (1967) *Pesticides and Pollution,* New Naturalist Series, Collins, London and Glasgow.

Milner, C. and **Hughes, R. E.** (1968) *Methods for the Measurement of the Primary Production of Grassland,* International Biological Programme Handbook No. 6, Blackwell, Oxford.

Mueller-Dombois, D. and **Ellenberg, H.** (1974) *Aims and Methods of Vegetation Ecology,* J. Wiley & Sons, New York.

Newbould, P. (1970) *Methods for Estimating the Primary Production of Forests,* International Biological Programme Handbook No. 2, Blackwell Scientific Publications, Oxford.

New Zealand Forest Service (1970) *Conservation Policy and Practice,* New Zealand Forest Service, Wellington.

Ogino, K., Sabhasri, S. and **Shidei, T.** (1964) 'The estimation of the standing crop of the forest in northern Thailand', *The S.E. Asian Studies,* **4,** 89–97.

Orchard, B. (1961) 'An automatic device for measuring leaf area', *J. Exp. Bot.,* **12**(36), 458–64.

Ovington, J. D. (1965) *Woodlands,* English Universities Press Ltd, London.

Ovington, J. D. and **Madgwick, H. A. I.** (1957) 'Afforestation and soil reaction', *J. Soil Sci.,* **8,** 141–9.

Page, G. (1970) 'Quantitative site assessment: some practical applications in British forestry', *Forestry,* **43,** 45–56.

Patterson, S. S. (1956) *The Forest Area of the World and its Potential Productivity,* Dept. of Geography Publication, Royal University of Goteborg, Sweden.

Phillipson, J. (1970) *Ecological Energetics,* Arnold, London.

Penman, H. L. (1948) 'Natural evaporation from open water, bare soil and grass', *Proc. Roy. Soc. (Lond.) A.,* **193,** 120–45.

Pennington, W. (1969) *The History of British Vegetation,* English Universities Press Ltd, London.

Pimental, D. (1973) 'Food production and the energy crisis', *Science,* **182,** 443–9.

Polster, H. (1961) 'Neure ergebnisse auf dem gebiet der standortsokologischen assimilationsund transpirations forschung an forstgewechse', *Landwirtschaftwiss,* **10**(1), Deut. Akad Berlin: Sitzber.

Pyke, M. (1971) 'Novel sources of energy and protein', in Wareing, P. F. and Cooper, J. P. (eds), *Potential Crop Production,* pp. 202–12, Heinemann, London.

Reid, D. (1966) 'The impact of nitrogenous fertilizer applications on dry matter production of grasslands', *Proc. Xth Int. Grassland Congr.* Helsinki, pp. 209–13.

Renfrew, C. (1977) 'Ancient Europe is older than we thought', *National Geographic,* **152,** 615–23.

Rennie, P. J. (1962) 'Some long term effects of tree growth on soil productivity', *Empire Forest Review,* **41,** 209–13.

Russel, E. W. (1961) *Soil Conditions and Plant Growth,* Longmans Green & Co., London.

Russell, E. J. (1957) *The World of the Soil,* Collins, London & Glasgow.

Satoo, T. (1965) 'Production and distribution of dry matter in forest ecosystems', *Tokyo Univ. For.,* **16,** 1–15.

Schery, R. W. (1972) *Plants for Man,* Prentice Hall, New Jersey.

Schwanitz, D. (1967) *The Origin of Cultivated Plants,* Harvard University Press, Cambridge, Massachusetts.

Seddon, B. (1971) *Introduction to Biogeography,* Duckworth, London.

Simmons, I. G. (1974) *The Ecology of Natural Resources,* Edward Arnold & Co., London.

Slesser, M. (1975) 'Energy requirements of agriculture', in Lenihan, J. and Fletcher, W. W., *Food, Agriculture and Environment,* pp. 1–20, Blackie, Glasgow.

Slobodkin, L. B. (1959) 'Energetics in *Daphnia pulex* populations', *Ecology,* **40,** 232–43.

Soil Survey Staff (1960) *Soil Classification – A Comprehensive System – 7th Approximation,* U.S. Dept. Agriculture, Washington D.C.

Spedding, C. R. W. (1971) *Grassland Ecology,* Clarendon Press, Oxford.

Stark, N. (1972) 'Nutrient cycling pathways and litter fungi', *Bioscience,* **22,** 355–60.

Struik, G. J. (1967) 'Growth habits of Dandelion, Daisy, Catsear, and Hawkbit in some New Zealand grasslands', *N.Z. Journ. Agric. Research,* **10,** 331–44.

Tarrant, J. R. (1975) 'Maize: a new U.K. agricultural crop', *Area,* **7,** 175–9.

Tayler, R. S. (1965) 'The irrigation of grassland', *Outlook Agric.,* **4,** 234–42.

Taylor, J. A. (1958) 'The growing season', *Aberystwyth Memoranda* No. 1, U.C.W., Aberystwyth.

Teal, J. M. (1957) 'Community metabolism in a temperate cold spring', *Ecol. Mono.,* **27,** 283–302.

Troughton, A. (1957) 'The underground organs of herbage grasses', *Commonwealth Agricultural Bureaux,* Bulletin No. 44, Farnham Royal, England.

Walter, H. (1973) *Vegetation of the Earth,* English Universities Press Ltd, London.

Wareing, P. F. and **Phillips, I. D. J.** (1970) *The Control of Growth and Differentiation in Plants,* Pergamon Press, Oxford.

Watts, D. (1971) *Principles of Biogeography,* McGraw-Hill, London.

Weaver, J. E. (1958) 'Summary and interpretation of underground development in natural grassland communities', *Ecol. Monogr.,* **28,** 55–78.

Webber, P. J. (1974) 'Tundra primary productivity', in Ives, J. D. and Barry, R. G. (eds), *Arctic and Alpine Environments,* pp. 443–73, Methuen, London.

Weigert, R. G. and **Evans, F. C.** (1964) 'Primary production and the disappearance of dead vegetation on an old field in S.E. Michigan', *Ecology,* **45,** 49–63.

Welch, D. and **Rawes, M.** (1964) 'The early effects of excluding sheep from high level grasslands in the north Pennines', *J. App. Ecol.,* **1,** 281–300.

Whitmore, T. C. (1975) *Tropical Rain Forest of the Far East,* Clarendon Press, Oxford.

Whittaker, R. H. and **Likens, G. E.** (1973) 'Primary production: the biosphere and man', *Human Ecology,* **1,** 357–69.

Whittaker, R. H. and **Woodwell, G. M.** (1968) 'Dimension and production relations of trees and shrubs in the Brookhaven Forest, New York', *J. Ecol.,* **56,** 1–25.

Whittaker, R. H. and **Woodwell, G. M.** (1971) 'Measurement of net primary production of forests', in Duvigneaud, P. (ed.), *Productivity of Forest Ecosystems,* pp. 159–76, UNESCO, Paris.

Wielgolaski, F. E. (1972) 'Vegetation types and plant biomass in Tundra', *Arctic and Alpine Research,* **4**(4), 291–306.

Zuckerman, S. (1957) 'Forestry, agriculture and marginal land', Report by the *National Resources (Technical) Committee,* H.M.S.O., London.

INDEX

(References to diagrams are given in *italics*)